# Lecture Notes in Economics and Mathematical Systems

538

Founding Editors:

M. Beckmann
H. P. Künzi

Managing Editors:

Prof. Dr. G. Fandel
Fachbereich Wirtschaftswissenschaften
Fernuniversität Hagen
Feithstr. 140/AVZ II, 58084 Hagen, Germany

Prof. Dr. W. Trockel
Institut für Mathematische Wirtschaftsforschung (IMW)
Universität Bielefeld
Universitätsstr. 25, 33615 Bielefeld, Germany

Editorial Board:

A. Basile, A. Drexl, W. Güth, K. Inderfurth, W. Kürsten, U. Schittko

Springer
*Berlin*
*Heidelberg*
*New York*
*Hong Kong*
*London*
*Milan*
*Paris*
*Tokyo*

Stefan Kokot

# The Econometrics
# of Sequential Trade Models

## Theory and Applications
## Using High Frequency Data

 Springer

Author

Stefan Kokot
Am Schlagsbach 13
63303 Dreieich
Germany

Cataloging-in-Publication Data applied for
Bibliographic information published by Die Deutsche Bibliothek.
Die Deutsche Bibliothek lists this publication in the Deutsche Nationalbibliographie; detailed
bibliographic data is available in the Internet at <http://dnb.ddb.de>.

ISSN 0075-8442
ISBN 3-540-20814-3 Springer-Verlag Berlin Heidelberg New York

Springer-Verlag is a part of Springer Science+Business Media

springeronline.com

© Springer-Verlag Berlin Heidelberg 2004
Printed in Germany

Typesetting: Camera ready by author
Cover design: *Erich Kirchner*, Heidelberg

Printed on acid-free paper    55/3142/du    5 4 3 2 1 0

It does take maturity to realize that models are to be used but not to be believed.

- Henri Theil[1]

---

[1] Theil (1971), p. vi.

# Preface

The present study has been accepted as a doctoral thesis by the Department of Economics of the Johann Wolfgang Goethe-University in Frankfurt am Main. It grew out from my five year long participation in two research projects, "Econometric analysis of transaction intensity and volatility on financial markets", and "Microstructure on financial markets", that were both conducted by the chair of Statistics and Econometrics (Empirical Economic Research) at the Department of Economics and Business Administration, Johann Wolfgang Goethe-University in Frankfurt am Main and financed by the state of Hessen.

During this time I have benefitted from many people. First and foremost I would like to thank my thesis supervisor, Prof. Dr. Reinhard Hujer, for initiating and supporting my studies with great encouragement. I am also very grateful to Prof. Dr. Christian Schlag for acting as the second thesis supervisor. Furthermore, I wish to thank Prof. Dr. Joachim Grammig who introduced me to the topics covered in this study in the first place and helped me to sharpen my views on econometrics and financial market microstructure theory through many discussions and also through his willingness to work with me on several related studies.

I also want to thank my former colleagues at the chair of Statistics and Econometrics Dr. Kai-Oliver Maurer and Dr. Marc Wellner. I have worked with both of them on several working papers and have greatly benefitted from their outstanding wisdom and experience. The same holds true for my present colleagues, especially for Sandra Vuletić, who worked with me on a paper that is most intimately related to the topics of the present study. In addition, I have to thank her and Dr. Marc Wellner for proof reading the whole manuscript with great care. Thanks also go to Marco Caliendo, Dubravko Radić, Stephan Thomsen, Christopher Zeiss, Irene Kisters-Ostheimer, who sadly passed away, and Birgit Andres-Kreiner for providing me an environment, in which I was able to work hard, live fast, and sometimes laugh until coffee came running out my nose.

Furthermore I want to thank all our student assistants at the chair, especially Filiz Polat, Paulo Rodrigues, Daniel Ueberall, and Oliver Wünsche, who lend me a hand for many rather unpleasant tasks ("Oh no, not your *#!&? graphics again..."). Before I started to work on my doctoral thesis, I had the great pleasure to participate on a research project entitled "Long-Term Care in the EU", financed by the Volkswagenstiftung and conducted by the chair of Economic Policy and Labor Economics, and I want to thank Prof. Dr. Roland Eisen, and Hans-Christian Mager for their close cooperation.

Finally, I wish to express my deepest gratitude to the whole Kokot-clan (German section) for always supporting me enthusiastically, even though I started to behave somewhat *strange* during the course of this project. I am especially indebted to my mother Tereza, my father Dragutin, my sister Irene, and to Thaja Kokot for their love and guidance throughout my whole life. Without you, none of this would have been possible.

Dreieich, November 2003                                        *Stefan Kokot*

# Contents

Contents .................................................... IX

1  Introduction ............................................. 1

2  Trading Mechanisms on Financial Markets ................. 5
   2.1  Typology of Security Markets ......................... 5
   2.2  Market Participants and Institutional Setup on the NYSE .... 9
        2.2.1  Market Participants ............................ 9
        2.2.2  Handling of Orders and Execution ................ 10
        2.2.3  Order Routing and Information Systems ........... 13

3  Sequential Trade Models ................................. 15
   3.1  Market Microstructure Theory ......................... 15
   3.2  Microstructure Models of the Black Box under Asymmetric
        Information ......................................... 17
        3.2.1  Sequential Trade Models ........................ 17
        3.2.2  Walrasian Batch Models ......................... 20
        3.2.3  Critical Assessment ............................ 22
   3.3  The Basic Sequential Trade Model ..................... 24
   3.4  Extensions .......................................... 27
        3.4.1  Trade Size Effects, No-Trading Events, and History
               Dependence ................................... 27
        3.4.2  Discriminating Between Market and Limit Orders ..... 33
        3.4.3  Models for Dually Listed Assets ................. 37
   3.5  Estimation of Structural Models ...................... 43
        3.5.1  Estimation of the Basic Model Using Information on
               Buys and Sells ................................ 43
        3.5.2  Estimation of the Basic Model Using Information on
               Trades ....................................... 44
        3.5.3  Estimation of Related Models .................... 47

3.6   Results of Previous Studies ............................... 50

**4   Econometric Analysis of Sequential Trade Models** ........ 61
4.1   The EKOP Model and Finite Mixture Models.............. 61
    4.1.1   Motivation ........................................ 61
    4.1.2   An Alternative Version of the EKOP Model .......... 64
    4.1.3   A Multivariate Finite Mixture Poisson Regression Model 67
    4.1.4   A Mixture Regression Model Based on the Negative
            Binomial Distribution............................. 71
    4.1.5   Accounting for Intraday Seasonality ................ 73
    4.1.6   Autoregressive Specification of the Conditional Mean
            Function ......................................... 74
    4.1.7   A Markov Switching Approach ..................... 76
4.2   Model Evaluation and Specification Testing ................ 82
    4.2.1   Specification Tests in Static Mixture and Markov
            Switching Models ................................. 82
    4.2.2   Determining the Number of Regimes................ 84
    4.2.3   A Conditional Moment Test for Goodness of Fit....... 86
    4.2.4   Testing Parameter Restrictions..................... 87
    4.2.5   Testing for Autocorrelation ....................... 89
4.3   Mixture and Regime Switching Models in Econometrics ...... 91

**5   Empirical Results**.................................... 93
5.1   The TAQ Database ...................................... 93
5.2   The Trade Direction ................................... 96
    5.2.1   Algorithms for the Determination of the Trade Direction 96
    5.2.2   Empirical Evidence on the Accuracy of Classification .. 98
    5.2.3   Classification of Trades............................ 103
5.3   Descriptive Statistics.................................. 106
5.4   Estimation Results ..................................... 112
    5.4.1   Model Selection .................................. 112
    5.4.2   Parameter Estimates.............................. 120
    5.4.3   Specification Tests................................ 129
    5.4.4   Classification of Regimes .......................... 135
    5.4.5   Testing Parameter Restrictions..................... 142

**6   Conclusions**........................................... 145

**Appendix**................................................ 147
A.1   The Poisson Process .................................... 149
A.2   Maximum Likelihood Estimation of a Multivariate Poisson
    Mixture Model ......................................... 151
A.3   The EM-Algorithm .................................... 153
A.4   The Poisson Regression Model ........................... 156
A.5   The Negative Binomial Regression Model.................. 157

A.6 Moments of Mixture Distributions ........................ 160
A.7 Unobserved Individual Variation of Trade Arrival Rates ...... 167
A.8 Markov Chains ......................................... 169
A.9 The Smoothing Algorithm .............................. 172
A.10 Estimation of Transition Probabilities in the Markov
      Switching Model ...................................... 173
A.11 Moments of the Dependent Variable in a Markov Switching
      Model ................................................ 174

**References** .................................................. 177

**List of Figures** ............................................. 189

**List of Tables** ............................................. 191

# 1

# Introduction

Today it is customary that every single transaction of a financial asset traded on major financial markets around the world is recorded electronically with detailed information about the time of occurrence, price and volume and other relevant characteristics. Recently, many of these so called *high frequency* data sets have become available at relatively low cost to academic researchers. Hence, the last fifteen years saw an unprecedented upsurge in both theoretical and empirical work related to the analysis of *market microstructure* issues using transaction data sets that are steadily increasing in size.[1] This upsurge went hand in hand with an equally unprecedented progress in computer technology, that made empirical analysis of huge data sets with ordinary desktop computers possible. It seems only natural then, that this innovation in both, the quality of data available for research and the development of the relevant economic theory was accompanied by the introduction of econometric methods which were tailor-made for the analysis of many policy issues related to market microstructure, e.g. how trading mechanisms should be structured in order to enhance market viability, or how parallel trading of a security on several markets affects prices.

One of the most promising new approaches, that brings together both econometric analysis and economic theory is the *sequential trade model*, initially introduced as a purely theoretical framework for the analysis of informational aspects of the price discovery process on financial markets by Copeland and Galai (1983), Glosten and Milgrom (1985) and Easley and O'Hara (1987). Sequential trade models focus on the dynamic interaction of three different types of economic agents on a security market, a market maker who supplies liquidity to the market by continuously posting prices at which the security may be bought and sold, and two types of traders, who differ with respect to their information set on future price movements. This framework allows one

---

[1] See e.g. Easley and O'Hara (1995), O'Hara (1995), Hasbrouck (1996), Goodhart and O'Hara (1997), Madhavan (2000), or Stoll (2001) for reviews of this branch of literature.

to analyze how the market maker sets prices in an environment that is characterized by an inherent uncertainty about whether any trader has superior information about the fundamental value of the security. In a seminal paper, Easley, Kiefer, O'Hara, and Paperman (1996) introduced an econometric framework that can be used to estimate structural model parameters employing high frequency data sets. This stimulating paper turned out to initiate an abundance of contributions that extended their approach both through a refinement of the theoretical model and through extensions of their econometric approach in several directions.

Despite of the numerous and highly innovative work that has been subsequently published, the focus of these contributions was limited in several ways. It is our impression, that this unnecessary limitation in the first place stems from an reluctance to extend the empirical framework introduced by Easley, Kiefer, O'Hara, and Paperman (1996). Their estimation strategy was based on data sets of trade event counts obtained from daily sampling schemes. This sampling scheme has been taken over by virtually all of the subsequent empirical studies of sequential trade models, even though many of these contributions extended the economic model to analyze different policy issues. However, the use of data sets of daily observations precludes the analysis of many interesting and important empirical phenomena, that have been found recently in several explorative studies of financial high frequency data sets.[2]

Our intention is to contribute to this branch of literature by extending the scope of the econometric framework to the analysis of high frequency data sets based on intradaily sampling schemes. In order to achieve this goal, we introduce a reasonable and quite general statistical model for multivariate time series of trade event counts that can be used for forecasting purposes as well as for tests of the implications of economic theory. This will be achieved by elaborating the close connection between sequential trade models and the class of statistical mixture models explicitly.

Mixture models have a long tradition in the statistical and econometric literature, and proved to be a very flexible empirical framework in a multitude of empirical applications. As we will show, there are some interesting extensions of the economic theory, that arise quite naturally by interpreting the sequential trade model as a statistical mixture model and in turn, extensions of already established statistical estimation methods may be motivated by the economic theory of sequential trade models. Furthermore, this approach enables us to apply several statistical tests to assess the goodness of fit of our econometric model in a way that is unprecedent in previous studies in this field. Taken together, we believe that this promising marriage of modern economic theory and sophisticated statistical methods may lead to many new insights and hence expand our understanding of the process of asset trading in several ways.

---

[2] Compare e.g. Engle and Russell (1998), Hasbrouck (1999), or Grammig, Hujer, and Kokot (2002).

This study is structured as follows: Chapter 2 contains a brief review of the trading mechanisms on existing financial markets. We describe the most important institutional aspects of the trading process, with a special emphasis devoted to the existing regulatory framework on the New York Stock Exchange. This focus is justified because most of the empirical studies employing the sequential trade framework that we will review later, as well as our own subsequent empirical application is based on data sets collected from the New York Stock Exchange. Chapter 3 begins with a short introduction to the basic topics of market microstructure theory in Sect. 3.1. A comparison of the sequential trade model to alternative models that have been proposed to analyze informational aspects of the price discovery process on asset markets follows in Sect. 3.2. In Sect. 3.3 we discuss the basic sequential trade model introduced by Easley, Kiefer, O'Hara, and Paperman (1996) extensively, since this will be the point of departure for the extensions we develop later. We also provide a detailed review of several alternative types of sequential trade models, both in terms of their theoretical content (Sect. 3.4) and in terms of the econometric methods that have been in use so far (Sect. 3.5). The Chapter concludes with a summary of the main findings of previous empirical studies related to sequential trade models in Sect. 3.6.

The econometric framework employed in this study is introduced in Chap. 4. It starts with a motivation of our modus operandi based on empirical regularities found in earlier studies of high frequency financial market data sets and shows that similar features are also contained in the data sets employed in this study in Sect. 4.1.1. We introduce several extensions of the basic sequential trade model and discuss different estimation procedures for these models and their applicability in subsequent Sections. In Sect. 4.1.2 we introduce a generalized formulation of the basic model. The first major extension we propose in Sect. 4.1.3 allows us to incorporate additional explanatory variables in a multivariate Poisson mixture model. We present a related mixture regression model based on the negative binomial density and discuss the economic implications of this model in Sect. 4.1.4. In Sects. 4.1.5 and 4.1.6 we review methods to account for serial dependence and intraday seasonality in regression models for count data. Finally, in Sect. 4.1.7 we introduce a Markov regime switching regression model and discuss its economic content.

Section 4.2 is devoted to statistical inference. In Sect. 4.2.1 we point out some general difficulties of hypothesis testing that may arise when established test procedures are applied to mixture and regime switching models. Section 4.2.2 introduces statistical tools that may be used to determine the appropriate number of mixture components. We continue by reviewing a statistical test for the assessment of the goodness of fit in Sect. 4.2.3, tests of parameter restrictions in Sect. 4.2.4, and tests for residual autocorrelation in Sect. 4.2.5 and discuss, how they have to be modified so that they can be applied in our econometric framework. Chapter 4 closes with a brief comparison to related work on mixture and regime switching models in the econometric literature in Sect. 4.3.

Our empirical applications are presented in Chap. 5. In a first step we introduce the transaction data base in Sect. 5.1. Before we can start to apply our econometric framework, we have to conduct several steps of data preparation. The estimation of sequential trade models is based on time series of *signed trades*, id est we have to determine whether any observed trade was initiated by a buyer or a seller. This information is not included in our original data set, so we have to apply one of several algorithms designed for the determination of the trade direction. Since this appears to be a sensitive topic, we decided to discuss this step in greater detail in Sect. 5.2.

Section 5.3 presents descriptive statistics of the data set of trade event counts that we use in our main empirical analysis contained in Sect. 5.4. We start with a discussion of our model selection strategy in Sect. 5.4.1. Then we present parameter estimates in Sect. 5.4.2 and results of specification tests in Sect. 5.4.3. The interpretation of the estimation results in the light of the underlying sequential trade framework is carried out by characterizing the nature of the information regimes we identified in Sect. 5.4.4, and by testing parameter restrictions motivated by economic theory in Sect. 5.4.5.

In Sect. 6 we summarize our main results and give a perspective on possible issues for future research. We include an extensive Appendix, that collects short reviews of some basic models and methods we employed in the course of this inquiry. Most of this material can be found in standard statistical and econometric textbooks in greater detail, so we also provide the relevant references for the interested reader. The Appendix also contains mathematical derivations and proofs that we refer to on several occasions in the main text. We felt that doing so could help to enhance the legibility of the present study.

# 2

## Trading Mechanisms on Financial Markets

### 2.1 Typology of Security Markets

In this Section we will review the main features of financial exchanges, derive a simple typology of markets, and classify the trading process on the New York Stock exchange (NYSE) within this framework. From a purely economic viewpoint there are three basic trading mechanisms, namely *Walrasian auction markets*, *quote driven markets* and *order driven markets*.[1]

A *Walrasian auction market* is characterized by a tâtonnement process, which is supervised by a Walrasian auctioneer. This auctioneer collects incoming buy and sell orders prior to trading, aggregates them in order to obtain market demand and supply schedules and announces a provisional first price. Investors who wish to buy or sell are then allowed to revise their orders in the light of the price proposal of the auctioneer. After resubmission of their orders, the auctioneer again aggregates buy and sell orders and announces a new provisional price. This procedure is continued until at the end of the subsequent *tâtonnement process*, a market clearing price is found, which equates market demand and supply. Trading is then allowed to take place at this equilibrium price. The London Gold Fixing is probably the real world trading system that comes closest to a Walrasian auction market design.[2] Some financial markets have features that are more or less similar to a Walrasian auction, although they do not employ Walrasian auctions as an exclusive trading mechanism. Examples include the opening batch auctions at the NYSE and the Paris Bourse.

In a *quote driven system*[3] investors can obtain firm price quotations (*quotes*) prior to order submission. The essential feature of this type of trading system is the presence of a delegated *market maker*, usually an exchange offi-

---

[1] See Bauwens and Giot (2001), pp. 1-35.

[2] See O'Hara (1995), p. 7.

[3] These systems are also known as *price driven markets*, *dealership markets* or *market maker systems*.

cial or an employee of a bank that is affiliated with the exchange.[4] The market maker continuously quotes firm bid and ask prices at which he is obliged to trade upon request up to a given number of shares. The quoted maximal volume at which the market maker is willing to trade at his best bid respectively ask price is called the *depth* and may be set by the market maker at his own discretion.

The main objective for the delegation of a market maker is to enhance the provision of liquidity on a continuous basis throughout the trading day. Market makers provide liquidity by continuously posting firm quotes, thus standing ready to buy (at the actual *bid* price) or sell (at the actual *ask* price) the asset at any time during the regular trading interval. The market maker thus becomes the counterpart of every single transaction that takes place on the trading floor. This is in sharp contrast to the Walrasian auctioneer who does not take any trading position himself, but acts only to redirect quantities from sellers to buyers. Examples of purely quote driven systems include the National Association of Securities Dealers Automated Quotation system (NASDAQ), the Chicago Mercantile Exchange (CME), foreign exchange markets (FOREX) and the International Stock Exchange in London.

In an *order driven trading system*[5] investors submit their orders for execution in an auction process. Typically they enter buy or sell limit orders into an *open order book*[6] maintained by the exchange. Trades occur only, when orders in opposite directions (i.e. a buy and a sell order) can be matched, i.e. the submitted prices of both trades correspond to each others. If the quantities (volumes) of the orders disagree, the order with the higher volume is only partially executed and the unresolved part of this order remains in the order book until it can be filled against other market or limit orders or it is cancelled from the order book upon request by the investor or because of expiry of its validation (usually at the end of the trading day during which it has been submitted).[7]

Thus, the absence of a delegated market maker, who provides liquidity by setting firm quotes, distinguishes order driven trading mechanisms from quote

---

[4] In practice market makers may be allowed to handle more than one security in a quote driven system.

[5] Order driven systems are sometimes also called *auction markets* or *order book markets*.

[6] Thus the quantity and price entries in the order book are *common knowledge*, i.e. all traders are allowed to view the order book directly. This is in contrast to some existing trading systems as explained below in more detail.

[7] In some order driven trading systems, e.g. the former German electronic integrated trading and information system, better known as IBIS, the investor is allowed to declare his limit order as *indivisible*. In such a case, the limit order may only be executed if both prices are equal and the quantity of the counterpart order is at least as big as the indivisible limit order.

driven systems.[8] The Paris Bourse, the Brussels Stock Exchange, the Swiss Option and Financial Futures Exchange (SOFFEX), the Frankfurt Stock Exchange and most of the existing electronic trading systems may be classified as operating with an order driven type of trading mechanism.[9]

Another important feature of the trading mechanism is whether trading takes place on a *continuous* basis or whether transactions may be conducted only at some discrete points in time. The latter structure is known as a *periodic* trading system. A continuous system may be described by a sequence of bilateral transactions at (possibly) different prices. In contrast, in a periodic system (also known as *call auction* or *batch market*) a set of multilateral transactions takes place at the same time at one common price. The choice between continuous and periodic trading will in general hinge on several factors, including the liquidity of the particular asset.[10]

The NYSE is a hybrid market as it combines elements from all three types of trading systems, Walrasian auctions, quote and order driven markets. As noted above, the batch auction conducted each morning before continuous trading takes place bears a close resemblance to the Walrasian auction market design. The delegation of market makers, called *specialists* in the official NYSE regulatory, is characteristic for a quote driven system, while the existence of an order book containing limit orders entered by traders is typical for an order driven market. Although the specialist is not committed to reveal outstanding entries in the limit order book to traders, the NYSE specialist must reflect the highest bid and the lowest ask price posted in the limit order book when they are better than his or her own quotes. Thus, it is the case that although all trading goes through the hands of the specialist, only a fraction of these trades actually involves the specialist directly in the sense that they affect his personal inventory position of stocks and cash.

Many trades stem from the execution of one limit order against another. According to the NYSE Fact Book for 1996, specialists on the NYSE participated directly in 18% of the total share volume transacted, while the remaining 82% were provided by other traders e.g. via direct matching of two corresponding limit orders. In general, the specialists role is more important

---

[8] See Glosten (1994) and Handa and Schwartz (1996) for an analysis of the motivation of investors to submit limit orders rather than market orders.

[9] See Cohen, Maier, Schwartz, and Whitcomb (1986), Chap. 2 and Domowitz (1993) for a more detailed classification of existing trading systems.

[10] As shown by Madhavan (1992), a periodic call auction system can be expected to offer greater price efficiency and may function, where continuous trading mechanisms fail either because of insufficient liquidity or a high degree of information asymmetry among market participants. This comes at the cost of a loss of immediacy and higher costs of gathering market information, which is revealed by the sequence of price quotations in a quote driven trading environment. Another interesting result of Madhavan (1992) is that while the equilibrium prices in an order driven, periodic batch auction market and in a Walrasian market may be expected to be the same, the traded volume is higher in a Walrasian market.

for less frequently traded stocks, than for blue chips.[11] Thus, limit orders may have a significant influence on the prevailing bid-ask spread and other trade characteristics on the NYSE.[12] In a certain sense the specialist faces competition from the order book, since his capability to widen the spread between the bid and ask price is de facto limited by the system of limit order submission. Whether this limitation is binding in practice, depends on the extent to which liquidity is provided by limit order submission.

Furthermore the NYSE is an example of a *centralized market*, since traders gather at a single location, the trading floor.[13] In contrast, at a *decentralized market* traders may be located at geographically different places. Electronic trading systems typically have a decentralized market structure: Traders, usually employees of a bank or a brokerage company, are located at separated trading rooms and connected via a centralized computer system through which the information contained in the order book is disseminated in real time to all participants.

An example of a decentralized stock market is the National Association of Securities Dealers Automated Quotation system (NASDAQ). The NASDAQ is a global intranet providing brokers and dealers with price quotations. Unlike the NYSE where orders meet on a trading floor, NASDAQ orders are paired and executed on a computer network. Access to the NASDAQ system is organized in three levels: Level I service provides participants only with the best bid and ask quotes in a given security without identifying the counterpart trader. Level II service additionally identifies the counterpart. Level III service allows registered traders to enter their own quotes.[14]

---

[11] The participation rates quoted above are measured as the number of shares bought and sold by all NYSE specialists in 1996 divided by the total number of shares traded, which equaled 104.6 billion shares in that year. See NYSE (1996), p. 5 and p. 18.

[12] See Chung, Van Ness, and Van Ness (1999) for an empirical analysis of the competition between the specialist and limit order traders. They show that the observed U-shaped intraday pattern of the bid-ask spread on the NYSE largely reflects the intraday pattern established by the variation of limit order placements and their corresponding price limits throughout the trading day.

[13] Nevertheless, the NYSE provides additional facilities, that can be used by external traders to submit orders to the NYSE, see the discussion of the Intermarket Trading System (ITS) in Sect. 2.2.3.

[14] In a slight variation of the terminology introduced in this study, the official NYSE regulatory calls all level III registered traders on the NASDAQ system *market makers*.

## 2.2 Market Participants and Institutional Setup on the NYSE

### 2.2.1 Market Participants

From a functional perspective, there are four basic types of actors on the NYSE:[15] Customers, brokers, dealers and market makers. In the real world two or more of these functions may be performed by the same person at the same time. The features that distinguish them are their rights and duties on the trading floor and whether they trade on their own account or on behalf of some other party.

*Customers* submit orders to buy or sell an asset, usually to a broker who has a commission for trading on the NYSE. These orders may be contingent on various features of the trading process (*limit orders*) or they may be direct orders to transact immediately, regardless of the state of the market (*market orders*). Customers are usually not physically present at the trading floor. *The broker* is an agent who acts as an intermediary between buyers and sellers. Brokers do not trade on their own account, but act as conduits for their customers. Brokers charge a commission for this service from their customers. *Dealers* are individuals or firms who buy and sell stocks on their own account, thus acting as a principal rather than as an agent. The dealer's profit or loss is the difference between the price paid and the price received for the same security. Since the same individual or firm may act either as broker or dealer on the trading floor, the dealer must disclose that he trades on his own account.

Central to the New York Stock Exchange trading system is the *market maker*, called *specialist* in the official NYSE regulatory. Specialists manage the batch auction market at the open of the trading day and they act as market makers during the continuous trading phase in the specific securities allocated to them. Specialist units are independent companies in corporate or partnership structures with the NYSE Inc.

Although each stock traded on the NYSE has a single market maker, NYSE specialists may be assigned multiple stocks. The number of stocks traded by an individual specialist varies according to the total activity of the stocks. Specialist firms distribute the activity to balance the workload for each specialist. Most specialists manage between 5 and 10 stocks. A specialist managing one of the most active stocks would normally specialize in that stock and, perhaps one less active stock. At the end of 1996 there were 37 specialist firms with 451 specialists. Thus, on average each specialist was responsible for making the market for 6.5 shares.[16] Usually each specialist employs a number of clerks, who perform a number of auxiliary tasks associated with order handling and reporting of trades and quotes.

---

[15] See Schwartz (1988), pp. 51-54, Schwartz (1991), pp. 27-31, and O'Hara (1995), pp. 8-11.

[16] At the end of 1996 there were 2907 companies listed on the NYSE with a global Market capitalization of 9.2 trillion US-\$. Source: NYSE (1996), p. 6.

Specialists must maintain a fair, competitive, orderly and efficient market. This means that all customer orders should have an equal opportunity to interact and receive the best price. It also means that once continuous trading begins, a customer should be able to buy or sell a reasonable amount of stock close to the last sale. Therefore, a specialist should try to avoid large or unreasonable price variations between consecutive sales. On the trading floor, specialists perform three critical roles:

- *Specialist as auctioneer*: The specialist manages the opening batch auction and continually posts bid and ask quotes throughout the trading day. These quotes are disseminated electronically through the NYSE quote and other market data systems that transmit the information instantly worldwide.
- *Specialist as agent*: A specialist acts as an agent for all limit orders. In this role, the specialist assumes the same fiduciary responsibility as any other broker.
- *Specialist as principal*: In order to maintain continuous trading activity, specialists also have to trade on their own account. Their own trading activity is restricted by several obligations. The first is to place and execute all customer orders ahead of their own. In 1996, roughly four out of five transactions at the NYSE took place between customers, without the capital participation of the specialist.

While specialists do not supply all the liquidity for the market, or determine the ultimate price of a stock, they use their inventory of stocks and money to bridge temporary gaps in supply and demand and help reduce price volatility by cushioning price movements. The specialist thus supplies short-term liquidity for the market and serves to keep the trading process viable. Specialist performance in meeting their obligations is closely monitored by the NYSE.

## 2.2.2 Handling of Orders and Execution

The actual trading process on the NYSE is regulated by a large and complex set of compulsory order execution rules. These rules are motivated by fiduciary obligations that are needed to keep the trading process at work in a principal/agent environment such as the NYSE. The regulatory framework does also define a number of order qualifications and instructions, that can be used by customers to instruct floor brokers with their individual trading preferences. Based on these rules, one can differentiate between quite a large number of different types of trades, although in practice most orders are written either as market or limit orders.[17]

---

[17] See Schwartz (1988), pp. 45-51, Schwartz (1991), pp. 35-41, and Hasbrouck, Sofianos, and Sosebee (1993).

*Market orders* are requests to buy or sell at the current market price. In effect market orders to buy (sell) are executed at the lowest ask (highest bid) quotation, given the specified volume of the request. Unless the specified volume of shares is unusually high, these orders may be expected to be immediately executed upon submission to the trading floor. However, at the NYSE the market maker is allowed to 'stop' incoming market orders upon request. In such a case the market maker guarantees execution at the prevailing quote, while attempting to execute the order at a better price at a later point in time.

A *limit order* sets a limit on the price at which the order may be executed. In the case of a limit buy (sell) order the maximum (minimum) price at which the request may be executed is fixed. Typically, these price limits are at or below (above) the price of the last transaction for buy (sell) orders. Limit orders that do not match the prevailing quotes at the time of their submission and therefore cannot be immediately executed, are put in the *limit-order book*, a file of orders sequenced by price and time of arrival. On the NYSE, the entries on the limit order book are not publicly accessible, but are closed to all but the specialist (market maker) and floor officials. Only the best bid and ask quotes are publicly announced. These quotes may in fact not stem from the order book, but result from the market makers own quotes, if they are better than those kept in the book.

*Stop orders* are also kept in the limit order book. These are requests to buy (sell) at prices above (below) the current quotes. However, in contrast to ordinary limit orders, stop orders are being converted into market orders, once the pre-specified price threshold (*stop price*) is being reached. Thus, they may be executed at a worse price than the specified stop price. It is also possible to submit stop orders that are converted into limit orders when the stop price is reached (*stop limit orders*). In this case, the execution price of a stop limit buy (sell) order will always be at or below (above) the specified stop price.

Orders that represent an unusually large number of shares may be destabilizing the market, if they are represented in the same way to the market as small orders and the liquidity of the market is not sufficient to handle such a large order. Therefore, investors intending to submit *block orders* may negotiate order processing in the so called *upstairs market*, rather than bringing them directly to the trading floor. The upstairs market provides a different trading mechanism, involving the services of an upstairs market maker, also called *block trader*, who stands ready to form a syndicate of counterpart participants for the block order. Once a syndicate is assembled, the block order is passed back to the floor, so its execution can be confirmed and recorded. In the official NYSE regulatory orders representing more than 10,000 shares are termed as block orders. However, the number of orders that are actually upstairs facilitated is much smaller than the total number of block orders submitted to the NYSE.

Typically shares are traded in 'round lots' of one hundred units. *Odd-lot orders* are orders for either less than a round lot (e.g. for 60 shares), or they represent the non-round part of a larger order (e.g. the last 60 shares of an order for 560 shares). Odd-lot orders are automatically executed against the market makers inventory and are priced at the current best bid or ask.

Occasionally, floor brokers will come up with orders, that represent both sides, buy and sell of a potential transaction. Such orders are called *crossing orders* and are subject to a number of specialized execution rules, beyond the usual priority rules, that we will describe below. The regulatory framework requires brokers to make a public bid and ask on behalf of both sides of the crossing order, thus giving other traders a chance to 'break up' the cross. Since crossing orders are in effect two orders in the opposite direction with the same price, the broker is required to offer at a price one minimum variation better than his bid before he may proceed with execution. In practice, crossing orders may be executed inside or outside the prevailing spread. Furthermore, there are some special rules for large crossing orders, e.g. for block orders that have been arranged in the upstairs market.

In addition to the basic order types, there are a number of special qualifications and instructions that may be stated by any customer when submitting orders to the NYSE. These instructions may be useful to customers when trying to control the conditions of order execution. *Day orders* for example are automatically cancelled, if they cannot be executed at the same day, when they are submitted, while *good-til-canceled* (GTC) orders remain on the book until they are executed or withdrawn by the customer. An order with a *fill or kill* (FOK) instruction for a specified volume of shares will either be entirely filled when the quoted market depth allows it, or cancelled immediately, while *immediate or cancel* (IOC) orders may be executed partially, but the unexecuted part of the order is being cancelled. The *all or none* (AON) qualification is similar to FOK, since it does not allow for partial execution, but AON orders will not be cancelled, if immediate execution is impossible. *At the opening* (ATO) and *market-on-close* (MOC) are instructions that restrain the timing of execution. An ATO order may only be executed at the opening of the trading day and any unfilled portion will be cancelled, while MOC orders are to be executed as near to the end of the trading day as possible and require special trading procedures.

Furthermore, there are a number of trading priority rules that may affect the sequence in which orders contained in the limit order book may be executed. The *price priority rule* states, that buyers (sellers) posting higher (lower) prices have priority over buyers (sellers) posting lower (higher) prices. For orders having the same price the execution priority is determined by the *time priority rule*: the first order placed is the first to be executed. In addition, a *size priority rule* that gives preference to the order with the largest volume at the same price, is also in effect and applies if the volume of the oldest limit order in the book is not sufficient to fill the trade completely. However, there are several special rules, that specify exceptions from these priority rules. The most important of these in practice is the priority granted to *public orders*: Specialists and floor traders who handle customer orders as agents and at the same time trade on their own account as principals must place back their own demands to public orders at the same price.

### 2.2.3 Order Routing and Information Systems

In order to disseminate information about current market conditions, the NYSE is required to report information on trades and quotes to interested parties. Typically, such information is recorded in real time and disseminated through the *Consolidated Tape System* (CTS), which is an information system for trade information, and through the *Consolidated Quote System* (CQS), which is the counterpart system used for quote dissemination. Both systems are maintained by the Securities Industry Automation Corporation. At the NYSE it is the duty of the member that represents the seller, to ensure that trades have been reported to the CTS. Actually, information on trades and quotes is reported either by floor reporters, who are employees of the NYSE, or by some specialist clerk rather than by the seller. The CTS displays real time information on transaction prices and volume, while the CQS reports current best bid and ask prices, as well as the corresponding bid and ask volume. All information that is processed through these systems is recorded electronically on the so called 'audit trail' data file, that supports NYSE surveillance operations and can be used by member firms to resolve disputes. Furthermore, a part of the data collected on the audit file is available for sale to outside parties. Thus, the CTS and the CQS are the ultimate source for the high frequency data sets available for scientific use.[18]

There are three major electronic order processing systems that provide access to the NYSE trading floor:[19] The *SuperDot*[20] is the main electronic order processing system for the NYSE. The SuperDot system serves to route orders by NYSE member firms directly to the specialists post on the trading floor, and execution reports back to the member firms. The *OARS* (Opening automated report system) is used to submit orders for the opening batch auction only and is in fact a part of the SuperDot system. The *Intermarket trading system* (ITS) is an electronic network connecting the NYSE, the American stock exchange (AMEX), NASDAQ and five regional exchanges[21]. ITS serves to disseminate the best available quotes among the participating exchanges, where they are accessible to market makers and floor brokers. Incoming market and limit orders for ITS eligible securities may thus be executed at the posted ITS quotes at possibly better prices than on the market where they originally arrived.

---

[18] See Sect. 5.1 for more details on financial transaction data sets.

[19] See Schwartz (1991), pp. 31-35, and Hasbrouck, Sofianos, and Sosebee (1993), pp. 21-29.

[20] The abbreviation *Dot* stands for 'designated order turnaround'.

[21] These are the Boston, Cincinnati, Midwest, Pacific, and Philadelphia stock exchanges.

# 3

# Sequential Trade Models

## 3.1 Market Microstructure Theory

In general, *market microstructure theory* might be defined as "the study of the process and outcomes of exchanging assets under explicit rules" (Easley and O'Hara (1995), p. 357) or "the area of finance that is concerned with the process by which investors' latent demands are ultimately translated into transactions" (Madhavan (2000), pp. 205-206). Therefore, the market microstructure literature is concerned with the analysis of how specific features of the trading process like the existence of intermediaries (e.g. the stock specialist or brokers), or the environment in which trading takes place (e.g. trading at a centralized exchange involving physical presence of traders versus trading on some decentralized electronic trading system) affect the price formation process. In a certain sense, the Null hypothesis of market microstructure theory is that many details of existing trading mechanisms and institutions, which are usually neglected in economic models, may exert a strong impact on the outcomes of the price formation process, namely prices and quantities.

Alternatively, one might argue, that market microstructure theory is a branch of microeconomics that tries to enlighten the black box of price setting behavior out of equilibrium. Standard microeconomic analysis is based on the assumption, that prices are determined by the intersection of demand and supply curves for a particular good. Analysis then focuses on how the equilibrium outcomes change, as one or both of the curves shift due to exogenous factors but the question of how exactly equilibria are being attained has long been treated as a question of secondary importance. It was only recently, that economists started to analyze more closely the determinants of behavior out of equilibrium.[1]

Many market microstructure models can therefore be interpreted as attempts to examine more closely the relationship between idiosyncratic features of the trading mechanism and the outcomes of trading in terms of prices

---

[1] See Spulber (1996) and O'Hara (1995), pp. 1-6.

and quantities. Madhavan (2000) distinguishes in his recent review of the literature between four main categories of research in market microstructure theory:

1. *Price formation and price discovery*: This refers to research on the black box by which latent demands are translated into realized prices and volumes, and includes issues like the determinants of trading costs or the process of gradual incorporation of information into prices.
2. *Market structure and design issues*: This category is concerned with the question, how different trading rules or protocols affect the black box and hence the liquidity and the quality of an asset market.
3. *Information and disclosure*: This category focuses on how the ability of market participants to observe information about the trading process affects the black box (market transparency).
4. *Informational issues arising from the interface of market microstructure with other areas of finance*: There are many overlaps between the analysis of the black box to other areas of finance and economics, e.g. corporate finance, asset pricing, international finance, or macroeconomics.[2]

In this study, we will focus on models that are concerned primarily with the first category, but may also be used to gain insights into topics that would belong to any of the other categories. The main theme of these models is the analysis of the effects of *asymmetric information* among market participants on the outcomes of the trading process.[3]

The origin of this strand of literature is usually credited to a paper written by Jack Treynor and published under the pseudonym Bagehot (1971). This paper introduced in a purely non-technical style the important distinction between *liquidity motivated* traders (also known as *uninformed* or *noise* traders), who possess no special informational advantages and *informed* traders who have private information that will help them to predict future price movements more accurately. It is important to notice, that the concept of an informed trader is distinct from that of an insider, usually defined as a corporate officer with fiduciary obligations to their shareholders. The concept of informed traders is broader, and may include insiders as well as traders, who have better forecasts of future cash flows, because they engage in costly information gathering activities.

Uninformed traders might be smoothing their intertemporal consumption stream through portfolio adjustments or trade because of the occurrence of unpredictable events such as job promotions, unemployment, deaths or disabilities. Large financial institutions such as pension funds or insurance com-

---

[2] See O'Hara (1999) on the links between market microstructure theory and other lines of research in finance and economics.

[3] Alternative models that analyze the price formation process stress the importance of transaction and inventory costs or market power exhibited by a monopolistic market maker. See O'Hara (1995), Chap. 2 and more recently Stoll (2001) for an overview of this strand of the literature.

panies will typically be included in this category, since their trading activities are often driven by liquidity needs of their clients, risk hedging considerations, or portfolio-balancing. Therefore they are willing to accept a price for *immediacy of trade execution*, the loss to informed traders.[4] Whatever their motives for trading might be, they are treated as exogenous with respect to the formal models we will introduce later. Treynor suggested that the presence of traders with superior information may help to explain the behavior of market makers, since they may be expected to lose to informed traders on average, so they ought to offset these losses by gains from trades with uninformed traders. This suggests, that the bid-ask spread will contain an informational component as well.

Note also, that the distinction between public and private information is not related to their price effects in the first place, but to their effects on *trading activity*. A public information event will in general affect prices, but should generate little or no abnormal trading activity.[5] In contrast, private information events lead to abnormal trading activity that precedes price changes (and in fact causes them). If seemingly public information events affect trading activity, it is therefore reasonable to conclude that they have a private component as well. Easley and O'Hara (1995) distinguish between two general approaches to analyze these issues: *Sequential trade* and *Walrasian (batch)* models. Both approaches focus on informational aspects of the price discovery process. In the next Section, we will describe the foundations of both approaches and compare them to each other.

## 3.2 Microstructure Models of the Black Box under Asymmetric Information

### 3.2.1 Sequential Trade Models

The three classical papers that introduced formalized (mathematical) *sequential trade* models are the papers by Copeland and Galai (1983), Glosten and Milgrom (1985) and Easley and O'Hara (1987). Copeland and Galai (1983) derived a one period model of the market makers pricing problem, given that some fraction of traders have superior information on the assets true value. The market maker is risk neutral, either monopolistic or competitive, and sets quotes to maximize his expected profit. Uninformed traders know how the general pricing process works, but do not know the true value of the asset. Their motivation to trade is exogenous. The nature of the information process is left unspecified as well.

---

[4] As shown by Milgrom and Stokey (1982), if the uninformed trade for purely speculative reasons, it is optimal for them not to trade rather than to trade with a sure loss as a result of the presence of informed traders.

[5] This would indeed be the case, if asset markets work efficiently, in that all public information is incorporated into prices immediately.

The market maker cannot identify informed traders, they are anonymous, but he knows the probability that any given trade comes from an informed trader. All trades are for a fixed (unit) number of shares. All traders are risk neutral, but are allowed to have price elastic demand functions, so neither the informed nor the uninformed are forced to trade at the current bid and ask prices set by the market maker. Therefore the market maker is required to take price elasticities into account, when he sets quotes. He calculates the expected profits from any trade, taking into account the probabilities of losing to the informed and gaining from trades with uninformed. The quotes set by the market maker emerge from the solution to his profit maximization problem. When there is competition among market makers, each of them will have zero expected profits, otherwise some monopoly profits may occur.

The model by Copeland and Galai (1983) yielded some important insights into the nature of the price formation process, most notably that even in the absence of inventory holding and order processing costs a positive bid-ask spread may prevail. Nevertheless, their approach suffers from the static framework, which implies, that all private information is made public after a trade is conducted. Even when repeated rounds of trading are admitted, the expected loss of the market maker will remain constant and he therefore has no incentives whatsoever to change his quotes. In this framework the trading process itself is not informative at all. But if private information is not revealed after each trade, the expected loss of the market maker will also depend on how quickly his prices will reflect the true value of the asset. Thus, in a dynamic version of the market makers pricing problem, sequences of buys and sells can reveal the underlying information and affect the behavior of prices.

The implications of learning from the order flow have been analyzed in a dynamic framework by Glosten and Milgrom (1985) and Easley and O'Hara (1987). In a competitive market informed traders will reveal their information by selling the asset if they know *bad news*, i.e. that actual asset prices are higher than the fundamental value given the private information and buying when they observe *good news*, i.e. when the actual asset price is lower than the fundamental value. Since informed traders know, that prices will eventually converge to fundamental values once the information is revealed, they seek to trade before their information becomes public. Therefore, the fact that someone wants to sell may be interpreted as a signal to the market maker that bad news have been received by this trader, but it may also mean, that the trader is uninformed and simply needs liquidity. The genuine idea in the framework of Glosten and Milgrom (1985) is that the market maker will protect himself by adjusting his beliefs about the value of the asset, conditioning his price setting behavior on the types of trades he observes. The enduring uncertainty on the side of the market maker about the nature of private information provides an incentive to change his quotes successively. Glosten and Milgrom (1985) show that sequences of trades on one side of the market will enable the market maker to learn about the nature of private information and his prices will eventually converge to the expected value of the asset, given the information. This explicit linkage of the price setting behavior to the existence of private information was novel and opened a new door for the analysis of how actual trading mechanisms can affect the black box.

In Easley and O'Hara (1987) this approach was further refined with respect to the information process itself, and also with respect to characteristics of trades other than the mere direction. They introduced the notion of *information uncertainty*. In earlier work, it was assumed that even though the market maker cannot observe the information event, he is always aware, that there are traders who know either good news or bad news. Now a third possibility is introduced, the absence of any private information. In this case the market maker will receive orders only from uninformed traders.

Furthermore, in the framework of Glosten and Milgrom (1985) and Copeland and Galai (1983) it was assumed, that a fixed (unit) number of shares is traded in each transaction. Relaxing this assumption enables the market maker to learn from both the direction and the size of trades. In Easley and O'Hara (1987) two different trade sizes, *large* and *small* quantities may be traded, and it is assumed that uninformed traders will transact both quantities. Otherwise, trade size will identify the type of trader unambiguously and the market maker will set the quotes for large trades equal to the expected values of the asset conditional on the type of information event. In this case no informed trader can exploit his information advantage by trading large amounts, and therefore no large trades will be observed at all.

They derive expressions for the market makers price setting function and show, that in this environment his prices will converge as well to the expected true value of the asset given the information. In this setup two types of equilibrium may prevail, depending on the magnitudes of the exogenous parameters. In a (*semi-*)*separating* equilibrium informed traders will conduct only large trades to exploit their informational advantage, and the market maker will thus set a spread only for large trades. On the other hand, in a *pooling* equilibrium informed traders will trade both quantities and the market maker will set a larger spread for large trades than for small trades, a feature often observed in existing markets. Their approach yielded further insights into how the trading process affects price setting behavior, as they showed how the size of trades affects price discovery.

There exist several refinements of the sequential trade approach. Diamond and Verrecchia (1987) analyze the implications of short-selling restrictions in a model very similar in spirit to the Glosten and Milgrom (1985) framework. Both, John, Koticha, Narayanan, and Subrahmanyam (2000) and De Jong (2001) extend the Glosten and Milgrom (1985) model to investigate the impact of parallel trading in option and stock markets. In Easley and O'Hara (1991), the possibility of submitting limit orders was incorporated into the Easley and O'Hara (1987) framework, while Easley and O'Hara (1992a) examined the effects of trading volume on the reaction speed of price adjustments to information. In a subsequent paper Easley and O'Hara (1992b) extend this framework to the analysis of the role of time in the trading process. Saar (2001b) examines price impacts of block trades in a sequential trade model. The notion of *demand uncertainty* is introduced in two recent papers (Saar (2000) and Saar (2001a)) as another source of asymmetric information by incorporating additional heterogeneity among traders through differential preferences and endowments. None of these papers contains any empirical analysis however.

### 3.2.2 Walrasian Batch Models

The fundamental difference between sequential trade models, and the Walrasian batch auction approach is related to the role of the market maker in the trading process. In the Walrasian approach market makers observe the net order flow from traders and set a single price, at which all orders are being executed.[6] The trading process is modelled as a sequence of auctions based on orders that are simply requests to buy or sell a specified number of securities at the prevailing market price, hence they are not price contingent. Therefore these models do not allow to characterize the bid-ask spread, but rather focus on the effects of strategic order placement by informed (and uninformed) traders on prices.

The prototype Walrasian batch model was presented in a seminal paper by Kyle (1985).[7] In his model a single, risk neutral market maker trades with a group of uninformed noise traders and a single, risk neutral informed trader, who receives a signal on the true value of the asset. Trading takes place in two steps: In the first step uninformed and informed traders simultaneously decide on the quantities they want to trade. In step two, the market maker sets a single, market clearing price, given the observed, aggregate order flow. If it is necessary to avoid disequilibria, the market maker himself stands ready to take a position in the asset at each auction, thus providing the volume, that will make the market clear at the price he sets.

The order flow submitted by uninformed traders is a random variable with a normal distribution having zero mean and known variance. Their orders are unrelated to prices, trading history or information. The market maker is aware of the presence of the informed trader, but cannot identify him. He has a prior normal distribution on the assets true value which he will update after he observes the aggregate order flow from all traders. He acts competitively, setting the price equal to the expected value of the asset according to a pricing function, and thus will receive zero expected profits.

The informed trader has rational expectations, and selects an order strategy consisting of the quantity of his market order, that will allow him to

---

[6] The label *Walrasian* stems from the close resemblance to Walrasian auctioneers, who keep markets in equilibrium by managing the price setting process, although the market maker definitively does not play the same role as a Walrasian auctioneer, as we will see.

[7] See Kyle (1984) and Kyle (1989) for related work.

maximize his expected profits, given knowledge of the pricing function of the market maker and the distribution of uninformed order flow. An equilibrium in this market setting is defined by a price setting function used by the market maker and an order strategy used by the informed trader. Kyle shows, that an unique equilibrium exists, in which both, the pricing function and the order strategy are linear functions of the observable variables that define the information sets of the market maker and the informed trader.[8]

Kyle considers single period and multiperiod versions of this model. He shows, that the sequence of prices will follow a martingale process with constant volatility over time and will eventually reflect all new information received by the informed trader. Furthermore he shows, that the informed trader's expected profits are higher when he trades continuously, rather than trying to manipulate prices by some mixed strategy, involving no order submission in some auctions. However, the informed trader will change the size of his orders over time, thus attempting to hide from the market maker.

Though Kyles model shows in an elegant way, how an informed trader may exploit his informational advantage by strategic interaction with the market maker, and how in turn prices are affected, there are several shortcomings in his approach. Much of the subsequent research that followed therefore focused on characterizing the equilibrium, when some of the assumptions made by Kyle are being relaxed. The studies that were based on the batch auction approach and extended Kyles model in one way or the other, are too numerous to be enumerated here completely.[9] Instead we will only point to a few of the main directions, in which his model has been extended.

The assumption that the market maker sees only aggregated order flow, but not individual orders is highly abstract, given the central position of market makers in most real world trading systems. However, when the market maker sees individual orders, the ability of a single informed trader to earn profits on private information may be drastically reduced, since the market maker may identify him. Forster and George (1992) explore several variants of Kyles batch auction model, in which information on the extent of liquidity trading is revealed to the market maker[10] before order submission (and trading) takes place. They show, that public disclosure can decrease the transaction costs of liquidity traders and may have differential effects on the sensitivity of prices to private information and on the incentives for private

---

[8] The information set of the market maker involves the parameters of his prior distribution, the parameters of the distribution of uninformed order flows, and the realized aggregate order flow of uninformed and informed traders. The information set of the informed trader involves the parameters of both mentioned distributions plus the signal he has received, but not the realization of the uninformed order flow.

[9] See O'Hara (1995) for a comprehensive exposition.

[10] They also explore several cases of public disclosure on liquidity trades to other market participants as well.

information acquisition, depending on the nature of the information revealed about liquidity trades.

Furthermore they show, that if there are multiple informed traders, the competition between them will force prices almost immediately to full information levels, thus abolishing the possibility to earn profits on private information. The issue of competition between multiple agents with private information has also been explored by Holden and Subrahmanyam (1992), Spiegel and Subrahmanyam (1992), and Foster and Viswanathan (1993). Their findings cast some doubts on the robustness of the properties of the linear equilibrium in the Kyle model. Relaxing the normality assumption leads to similar results. In the Foster and Viswanathan (1993) model non-normality of the market maker's prior distribution results in an equilibrium in which information is fully revealed through prices, so that incentives to engage in costly information gathering activities are diminished.[11]

Another strand of the literature extends the Kyle model by allowing uninformed traders to act strategically too. This is achieved by letting them have discretion over the timing of their trades as in Admati and Pfleiderer (1988) and Admati and Pfleiderer (1989), or by assuming they trade to hedge the risk of exogenous endowment shocks as in Spiegel and Subrahmanyam (1992).[12] Furthermore the decision to engage in information gathering activities may be endogenous too.[13] In Admati and Pfleiderer (1988) and Admati and Pfleiderer (1989), the interaction of optimizing decision making by uninformed and informed traders produces patterns in intradaily trading activity, which tends to cluster near the open and the close. The concentration of trades results from increased order submission by both types of traders.[14]

### 3.2.3 Critical Assessment

We encountered two different approaches to modelling security market behavior in the presence of asymmetric information. There are several critical aspects to both approaches, that we have to address explicitly.

The sequential trade approach has been criticized for several reasons.[15] Although it is possible to show that prices converge to full information values, questions about the speed of adjustment are not so easily resolved in the sequential trade framework. Sequential trade models implicitly assume that traders form a queue and after they have traded once, each trader rejoins at the end of the queue, if he wants to trade further. This may be at odds with real behavior, especially with respect to informed traders, who may want to

---

[11] See Rochet and Vila (1994) for related work.

[12] See also Foster and Viswanathan (1990) and Seppi (1990).

[13] See Verrecchia (1982) for an early attempt to tackle this issue and Foster and Viswanathan (1993) for related work.

[14] See Brock and Kleidon (1992) for an alternative explanation of intraday patterns in trading activity.

[15] See O'Hara (1995), pp. 73-74 and Easley and O'Hara (1995).

trade repeatedly in order to exploit their information advantages. The arrival rates of both informed and uninformed traders may not be constant in time. The analysis of strategic behavior both by uninformed and informed traders is very difficult to conduct. Implementing such behavior in a sequential trade framework may result in very complex models. Furthermore, in these models there is no role for inventory effects or order processing costs that market makers face in practice as well.

The last point also holds true in Walrasian batch auction models. However, there are several other critical points to be made as well: In a batch-clearing system, trades occur at a single, market clearing price, while in actual markets, the bid-ask spread plays an important role. No explicit explanation of the dynamic process of determining quotes is given in batch auction models. Information revealed by no-trading intervals is obliterated by the assumed aggregation of trades. Furthermore, the effect of information uncertainty is not analyzed in these models, as it is assumed, that there will always be informed traders present in the market.

Thus, both types of models have their deficiencies and their merits. While Walrasian batch models allow to characterize the strategies of traders in different market settings, they do not allow any predictions on the behavior of bid-ask spreads, since these simply do not exist in this framework. On the other hand, sequential trade models allow to characterize the spread, but it is very hard to incorporate aspects of strategic behavior in this setting. Therefore, these two approaches should be best viewed as being complementary to each other, since they aim to explain different aspects of the effects of private information on the price formation and discovery process.

From an empirical point of view however, the sequential trade approach has one prime advantage over the Walrasian batch model approach. It represents not only a theoretical model of the effects of information asymmetries on price setting behavior, but it also provides a coherent framework that can be used to estimate structural parameters of the model from observable quantities that are typically contained in high frequency transaction data sets from financial markets. The Walrasian approach does not provide such a coherent framework, although many of the implications of these models have been tested in an abundance of empirical studies.[16]

In the next Section we will describe the sequential trade model introduced by Easley, Kiefer, O'Hara, and Paperman (1996) in greater detail. Their model may be viewed as a simplified version of its predecessors [Easley and O'Hara (1987) and Easley and O'Hara (1992b)], but since their framework captures all the essential features of sequential trade models and many other models may be viewed as generalizations of this particulary simple one, it will serve both as an appetizer to more complex models, that we will discuss later on and as the

---

[16] See Hasbrouck (1996), Goodhart and O'Hara (1997) and Coughenour and Shastri (1999) for surveys on the methods and results of empirical studies related to market microstructure issues.

starting point for econometric analysis. We will then turn to other variants of sequential trade models, that have been used as a basis for empirical studies.

## 3.3 The Basic Sequential Trade Model

In the framework of Easley, Kiefer, O'Hara, and Paperman (1996), henceforth denoted as EKOP, differences in the magnitude of the bid-ask spread between liquid and illiquid stocks are explained by differences in the frequency of occurrence of particular information events. Their setup is a mixed (discrete and continuous time) sequential model of the trading process, in which trades arise because of the interaction of three types of economic agents, *informed* and *uninformed traders* and a risk neutral, competitive *market maker*.

The magnitude of the bid-ask spread $\Upsilon(\tau)$ at any time $\tau$ during the regular trading day, $\tau \in [\underline{\tau}, \overline{\tau}]$, where $\underline{\tau}$ denotes the time of day at which trading begins and $\overline{\tau}$ the time of day when trading stops, depends on the arrival rates of informed and uninformed traders (which are governed by independent Poisson processes) and on the likelihood of the occurrence of three different types of information events (*no news*, *good news* and *bad news*) which are chosen by nature every day before the first trade takes place. Given the occurrence of a news event, which has probability $\alpha$, a good news event occurs with probability $1 - \delta$.

On a trading day without news events, all transactions that are carried out result from the arrival of buy and sell orders from uninformed traders. The arrival rate of both buy and sell orders is assumed to be determined by an independent Poisson process with arrival rate per minute of the trading day equal to $\varepsilon$.[17] All incoming orders are market orders, i.e. they are immediately executed at the prevailing bid and ask prices.

If a news event occurs, there will be additional order arrivals resulting from the transaction demand by informed traders, who are assumed to be risk neutral and competitive. Informed traders observe a signal, indicating either the presence of good or bad news, so their trade arrival rate will dependent upon the type of information event. The signal they observe is the true (fundamental) value of the stock at the end of the day given the information signal. Thus, if they observe e.g. a low signal indicating bad news, informed traders know that the fundamental value is equal to $\underline{V}_d$, and that prices will converge to this value. Therefore, when the actual ask price is higher than $\underline{V}_d$, the profit maximizing investment strategy will be to sell the asset, so the sell arrival rate will be higher than on a no news day. For similar reasons on a good news day there will be a higher rate of buys. EKOP assume that two independent Poisson processes govern the arrival for informed buyers and sellers, both having arrival rate equal to $\mu$. An additional independence assumption between the order arrival processes of informed and uninformed traders is required.

---

[17] See Appendix A.1.

The probability structure of the trading process is summarized by the tree diagram in Fig. 3.1.

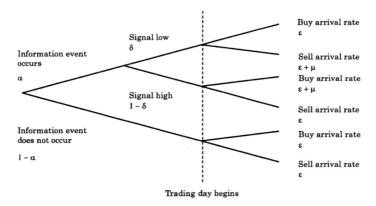

**Fig. 3.1.** The Easley, Kiefer, O'Hara and Paperman (1996) model.

Since the market maker cannot observe directly which type of information event has occurred, he uses Bayes rule[18] to update his beliefs about the nature of the information event throughout the trading day. In order to protect himself from losses due to the arrival of informed traders, he updates quotes so that his expected loss from informed trading conditional on the daily history of incoming buy and sell orders is equal to zero. Denote the market maker's subjective probability at time $\tau$ that a good news event has occurred by $p_g(\tau)$. The corresponding probabilities for bad news and no news events are given by $p_b(\tau)$ and $p_n(\tau)$. Let $S_\tau$ ($B_\tau$) denote the arrival of a sell (buy) order at time $\tau$. Using Bayes rule, the market maker's beliefs on the probabilities of the three types of news events conditioned on the sequence of incoming orders at time $\tau$ can be calculated, starting from his unconditional prior beliefs at time $\underline{\tau}$, which are equal to the probabilities, with which nature chooses the information regime

---

[18] *Bayes rule* may be interpreted as an updating formula, that can be used to form an inference on the subjective likelihood or *posterior belief* $p(\text{event}|\text{data})$ on the occurrence of a probabilistic event, given the observation of some data and a *prior belief*, which is in the form of a probabilistic inference $p(\text{event})$ as well. In general terms, Bayes rule may be stated by the following equation

$$p(\text{event}|\text{data}) = \frac{p(\text{event}) \cdot p(\text{data}|\text{event})}{p(\text{data})}$$

$$= \frac{p(\text{event}) \cdot p(\text{data}|\text{event})}{p(\text{data}|\text{event}) \cdot p(\text{event}) + p(\text{data}|\text{not event}) \cdot p(\text{not event})}.$$

See O'Hara (1995), pp. 78-82.

$$p_g(\underline{\tau}) = \alpha \cdot (1 - \delta) \qquad (3.1)$$
$$p_b(\underline{\tau}) = \alpha \cdot \delta$$
$$p_n(\underline{\tau}) = 1 - \alpha.$$

After observing the first order arrival, posterior probabilities for all three events can be calculated recursively. Given that a sell order arrives at time $\tau$, these probabilities are given by

$$p_g(\tau \mid S_\tau) = \frac{p_g(\tau) \cdot \varepsilon}{\varepsilon + p_b(\tau) \cdot \mu} \qquad (3.2)$$

$$p_b(\tau \mid S_\tau) = \frac{p_b(\tau) \cdot (\varepsilon + \mu)}{\varepsilon + p_b(\tau) \cdot \mu}$$

$$p_n(\tau \mid S_\tau) = \frac{p_n(\tau) \cdot \varepsilon}{\varepsilon + p_b(\tau) \cdot \mu}.$$

where $\mu$ is the arrival rate of informed and $\varepsilon$ the arrival rate of uninformed traders. The corresponding probabilities in the case of a buy order arrival are

$$p_g(\tau \mid B_\tau) = \frac{p_g(\tau) \cdot (\varepsilon + \mu)}{\varepsilon + p_g(\tau) \cdot \mu} \qquad (3.3)$$

$$p_b(\tau \mid B_\tau) = \frac{p_b(\tau) \cdot \varepsilon}{\varepsilon + p_g(\tau) \cdot \mu}$$

$$p_n(\tau \mid B_\tau) = \frac{p_n(\tau) \cdot \varepsilon}{\varepsilon + p_g(\tau) \cdot \mu}.$$

The bid-ask spread is determined by the following relation

$$\Upsilon(\tau) = \frac{\mu \cdot p_g(\tau)}{\varepsilon + \mu \cdot p_g(\tau)} \left( \overline{V}_d - E\left[ V_d \mid \mathcal{F}_\tau \right] \right) \qquad (3.4)$$
$$+ \frac{\mu \cdot p_b(\tau)}{\varepsilon + \mu \cdot p_b(\tau)} \left( E\left[ V_d \mid \mathcal{F}_\tau \right] - \underline{V}_d \right),$$

where $V_d$ is the value of the asset at the end of day $d$, $\mathcal{F}_\tau$ denotes the information set of the market maker at time $\tau$, which consists of the observed sequence of buys and sells by that time, $\overline{V}_d$ is the asset's true value conditional on observing good news on day $d$ and $\underline{V}_d$ the corresponding true value given bad news. Thus, according to the EKOP model, the spread contains two components, that reflect the market maker's possible loss at time $\tau$ due to asymmetric information. The first term in (3.4) is equal to the probability of observing an information based buy order, multiplied by the expected loss of the market maker caused by such a transaction, while the second term reflects the expected loss caused by an information based sell in a symmetric way.

In this framework, the difference between bid and ask quotes arises solely because of asymmetric information of market participants about the occurrence of information events. Other components of the spread such as those caused by the market maker's reluctance to maintain large inventory imbalances, or by the exercise of market power by a monopolist market maker are neglected. Nevertheless, other components of the spread may be controlled for, by simply extending the left hand side of (3.4).

Another important quantity that can be derived from the EKOP model is the *probability of informed trading* $PIT(\tau)$. This quantity is equal to[19]

$$PIT(\tau) = \frac{\mu \cdot (1 - p_n(\tau))}{\mu \cdot (1 - p_n(\tau)) + 2 \cdot \varepsilon}.$$

The initial belief of the market maker on the probability of informed trading

$$PIT = \frac{\mu \cdot \alpha}{\mu \cdot \alpha + 2 \cdot \varepsilon}, \tag{3.5}$$

may be interpreted as the unconditional share of informed trading. This quantity is especially interesting in empirical applications, e.g. in order to compare the importance of informed trading for different stocks or for the same stock traded on different markets.

## 3.4 Extensions

### 3.4.1 Trade Size Effects, No-Trading Events, and History Dependence

While many sequential trade models are straightforward extensions of the EKOP model with trading in continuous time, some of the models presented in this Section are based on a slightly different model of the trading process. The theoretical foundations for the inclusion of trade size and time effects have been developed in Easley and O'Hara (1987) and Easley and O'Hara (1992b). In this framework the trading day is divided into discrete time intervals, where each interval is long enough to accommodate just one trade. This assumption makes it possible to define a third, *no-trading* event in addition to buys and sells, on which the EKOP model focuses. This is in contrast to the continuous trading framework used in the EKOP model, in which no-trade intervals may

---

[19] Note, that $PIT(\tau)$ is the ratio of the probability, that any kind of trade (buy or sell) is information based, to the probability of observing any trade:

$$PIT(\tau) \equiv p(\text{news} \mid S_\tau \cup B_\tau) = \frac{p[(S_\tau \cup B_\tau) \cap \text{news}]}{p(S_\tau \cup B_\tau)}$$

$$= \frac{p(S_\tau \mid \text{news}) + p(B_\tau \mid \text{news})}{p(S_\tau) + p(B_\tau)}$$

$$= \frac{\mu \cdot p_b(\tau) + \mu \cdot p_g(\tau)}{\varepsilon + \mu \cdot p_b(\tau) + \varepsilon + \mu \cdot p_g(\tau)}.$$

Replacing $p_b(\tau) + p_g(\tau)$ by $1 - p_n(\tau)$ yields the expression given above.

arise in the course of the trading day as well, but are not treated different from observations on buys and sells in empirical applications.[20] Through the introduction of the no-trade events, the duration between trades will have an effect on the market makers pricing behavior, because long durations between trades indicate the absence of information of any type. This provides an incentive to the market maker to lower the spread, when the intensity of trading is low. Note also, that the structural parameters that describe the behavior of informed and uninformed traders during the trading day in this framework have to be interpreted as probabilities, rather than arrival rates, which was appropriate in the continuous time setting of the basic model.

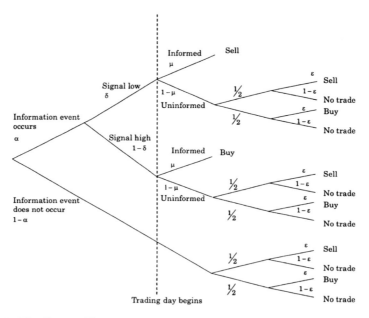

**Fig. 3.2.** The Easley, Kiefer and O'Hara (1997b) model without trade size effects.

Two different variants used for empirical applications of this model appear in Easley, Kiefer, and O'Hara (1997b). In the first model variant the effects

---

[20] In fact, no-trade intervals are treated implicitly as zero observations of the buy and sell sequences. Since most empirical applications are based on observations of the number of buys and sells *per trading day*, these zero observations are lost in the process of aggregation to daily time series in most cases. In practice, only extremely infrequently traded assets will occasionally have zero trades on some days. On the other hand, if estimation is based on intradaily observations of buys and sells for some smaller time interval, there will typically be many zero observations even for the most frequently traded assets.

of no trade events are considered. The structure of the information process is the same as in the EKOP model, so before trading begins, the type of the information event is chosen by nature. When an information event occurs, an informed trader is selected to trade with the market maker with probability $\mu$, while uninformed traders are selected with probability $1 - \mu$. If an uninformed trader is selected, he will equally likely be a *potential buyer* or a *potential seller* and he will decide to trade with probability $\varepsilon$ and not to trade with probability $1 - \varepsilon$, while informed traders will always trade, when they are selected. Trading continues throughout each trading day in the same fashion, thus giving rise to the probabilistic structure of the trading process depicted in Fig. 3.2.

In the second variant of this model, Easley, Kiefer, and O'Hara (1997b) additionally distinguish between two different trade sizes, *small* and *large* trades. In this variant, informed traders will purchase large quantities with probability $\gamma$ and small quantities with probability $1 - \gamma$, regardless of the type of signal they observed. Uninformed traders will buy (sell) large quantities with probability $\beta$ $(1 - \beta)$. This model variant nests the first variant, since if $\gamma = \beta$, trade size will convey no information on the state of the information regime, while when $\gamma > \beta$ large trades have more information content than small trades. Otherwise, the size of the trade provides no additional information to the market maker beyond that conveyed in the direction of the trade. The probabilistic structure of the second model variant is given in Fig. 3.3.

The framework of Easley, Kiefer, and O'Hara (1997a) is an extension of the second model variant of Easley, Kiefer, and O'Hara (1997b) in which the effects of trade size are incorporated into the sequential trade model. In addition their model allows to test for the possibility, that uninformed traders condition their trades on the observed order flow, thus inducing serial correlation in the observed trading process. They argue that it is unlikely, that uninformed traders will ignore the trading history when placing their orders as they are able to observe directly any information that makes the trading history useful for the market maker.

In their approach therefore additional parameters are introduced into the basic setup. The structure of the information process is the same as before. If a news event occurs on any particular trading day, informed traders arrive with probability $\mu$ and conduct a large trade with probability $\gamma$. For reasons of simplification, Easley, Kiefer, and O'Hara (1997a) assume that there are only two possible trade sizes (large and small trades) and that the size of the trade does not vary for buy and sell orders, so a small trade will be conducted by an informed trader with probability $1 - \gamma$, regardless of the type of information signal he observes. Analogously, uninformed traders will conduct a large trade with probability $\beta$ and a small trade with probability $1 - \beta$, regardless of the direction of their trades.

In contrast to the models introduced previously, uninformed traders will condition their decision to trade upon the observed order flow too. This is reflected by the introduction of additional history dependent parameters into the framework. With probability $\varepsilon_\tau$ an uninformed trade will occur at time

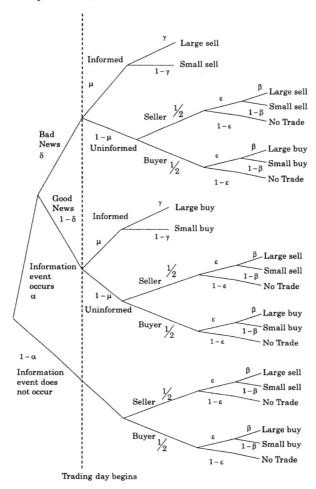

**Fig. 3.3.** The Easley, Kiefer and O'Hara (1997b) with trade size effects.

$\tau$. The probability of uninformed trading will dependent on the direction of the last observed trade during a predefined period of time prior to $\tau$. If there was no trade during the pre-trade interval, uninformed trades will occur with probability $\varepsilon(NT)$.[21] On the other hand, if the previous trade within the pre-trade interval was a buy then uninformed trades will occur with probability $\varepsilon(B)$ and with probability $\varepsilon(S)$ if it was a sell. Note, that only the direction of the previous trade matters, not the size, so there are only three different values that $\varepsilon_\tau$ may assume.

---

[21] In their empirical application they consider a five minute interval before each observed trade.

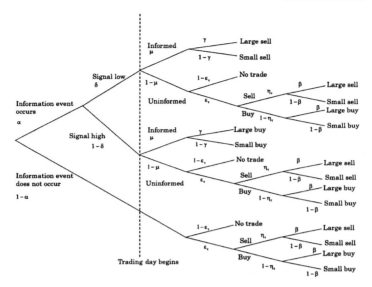

**Fig. 3.4.** The Easley, Kiefer and O'Hara (1997a) model.

The same structure is assumed for the determination of the trade direction of the uninformed. If the previous trade was a buy (sell), then the probability of an uninformed sell will be equal to $\eta(B)$ ($\eta(S)$), while a no-trade interval will be followed by a sell with probability $\eta(NT)$. Note, that the stochastic process of uninformed trade arrivals in this model follows a five state Markov chain with restrictions imposed on the transition probabilities. The states of this Markov chain are defined by the observable characteristics of the trades, so the five types of trading events are 'no trade', 'large sell', 'small sell', 'large buy', and 'small buy'. If e.g. the last observed trade before time $\tau$ was of the no trade type, then the transition probability for another no trade is equal to $1-\varepsilon(NT)$, while a large sell will be observed with probability $\varepsilon(NT)\cdot\beta\cdot\eta(NT)$. The transition probabilities for the informed traders are independent of the observed characteristics of trades, but depend only on the type of information event. The probabilistic structure of the model is summarized by the tree diagram depicted in Fig. 3.4.

The assumption of history dependence in the arrival rates of uninformed traders leads to quite complex expressions for the posterior probabilities of the market makers beliefs on the nature of the information regime, as the market maker has to keep track of all five observable types of trades, when conducting his updates. Nevertheless, the derivation of the corresponding expressions is based on the assumption, that the market maker is risk neutral, competitive and acts as a Bayesian learner and is thus analogous in structure to the EKOP model. This holds true for the two model variants in Easley, Kiefer, and O'Hara (1997b) as well. An alternative approach to uninformed trading behavior is introduced in Lei and Wu (2001) and Easley, Engle, O'Hara, and Wu (2002), who extend the EKOP model by allowing both informed and uninformed trade arrival rates to depend on past trading activity and additional exogenous variables.

In a related paper, Weston (2001) distinguishes between three types of traders: *informed traders*, who trade for speculative motives and observe a precise signal on the future value of the asset, *liquidity traders*, who trade for non-speculative motives as discussed in Sect. 3.1 and are aware of the fact that they do not observe any signal, and *noise traders*, who trade for speculative reasons, do not observe the signal, but believe they do. This group of traders receives a pseudo-signal, indicating that some information event has occurred, but not whether the fundamental value of the asset will rise or fall. Nevertheless, noise traders act as if their signal was perfectly correlated with future prices, thus increasing the total order flow on an information event day. The model is designed to explain the volatile behavior of trading volume, and the introduction of noise traders may be associated with the recent launch of day-trading firms, that allow individual investors to submit orders directly to stock markets.

In an extension of this model, noise traders receive a clean signal of whether an information event has occurred with probability $\phi$ and no signal with probability $1 - \phi$. However, the signal received by noise traders indicates only that some information event has occurred, but not whether it was good or bad news. Therefore, when they receive the signal, buy and sell orders submitted by noise traders arrive at the market with rate equal to $\varepsilon_N$, while liquidity traders always have arrival rates equal to $\varepsilon_L$, and informed traders either buy or sell, depending on the type of signal they receive, with order arrival rate equal to $\mu$. The structure of this model variant is given in Fig. 3.5. Note that this structure gives rise to six different information regimes, instead of three as in most other sequential trade models.

The noise trader model adds additional empirical content to the sequential trade framework, since it allows one to derive expressions for the probability of informed trading $(PIT)$ as well as for the probability of noise trading $(PNT)$. In the extended variant of this model, $PIT$ is given by

$$PIT = \frac{\mu \cdot \alpha}{\mu \cdot \alpha + 2 \cdot \varepsilon_L + 2 \cdot \varepsilon_N \cdot \Theta}, \tag{3.6}$$

while the implied $PNT$ is equal to

$$PNT = \frac{2 \cdot \varepsilon_N \cdot \Theta}{\mu \cdot \alpha + 2 \cdot \varepsilon_L + 2 \cdot \varepsilon_N \cdot \Theta}, \tag{3.7}$$

with $\Theta = (1 - \alpha) \cdot (1 - \phi) + \alpha \cdot \phi$. These quantities may be used to assess the relevance of noise trading behavior across different markets as well as for different stocks in empirical applications.

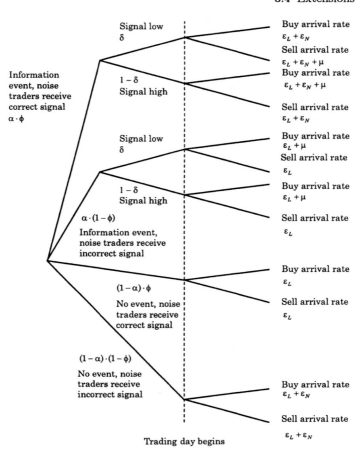

**Fig. 3.5.** The Weston (2001) model.

### 3.4.2 Discriminating Between Market and Limit Orders

Another important characteristic of many real world asset markets, in particular of stock markets, is the possibility to submit *limit orders*, which are allowed to be price-contingent, in the sense that their execution is delayed, until the prices set by the market maker match the demands of the limit order that is kept in the order book until execution. The variants of sequential trade models we encountered until now all abstracted from this feature, by assuming that all incoming orders are *market orders*, which are executed immediately by the market maker at the prevailing best bid or ask price available. Easley and O'Hara (1991) consider a sequential trade model variant, in which *stop orders* may also be submitted.[22]

Consider for example that a good news signal is observed by an informed trader. If the actual asset price is below the fundamental value at the end of the day, the informed trader knows, that prices will rise. When he submits a stop-loss order, he may expect, that it will never be executed, while a stop-buy order may only be executed at a higher price, than the actual. Thus, the profit maximizing strategy of the informed trader is to trade at the current price. Since analogous reasoning holds also in the case of bad news, informed traders will always choose to execute market orders rather than stop orders. The Easley and O'Hara (1991) model was not used to test this implication empirically however, and their focus is on the effects of order disclosure on the properties of the price process.[23]

Easley, O'Hara, and Saar (2001) developed a variant of the basic sequential trade model, in which uninformed traders may submit limit orders as well as market orders, while informed traders always conduct market orders. Thus, while informed traders arrive with rate $\mu$ and buy or sell immediately according to the type of information they observe, uninformed traders may belong to one of four different types (see Fig. 3.6): Uninformed buyers, who arrive with rate $\varepsilon_B$ may either submit market orders ($\kappa_B$ is the corresponding fraction of uninformed buyers ) or limit orders, which are executed with rate $\nu_I$ if they are in the same direction as the market orders of informed traders (e.g. buys on a good news day) and with rate $\nu_N$ if they are in the opposite direction (e.g. buys on a bad news day). Uninformed sellers arrive with rate $\varepsilon_S$, and $\kappa_S$ is the corresponding fraction of sellers who submit market orders. The rate of execution again depends on whether they are in the same direction as informed orders or not.

The introduction of the limit order execution rates $\nu_I$ and $\nu_N$ causes the likelihood of limit order execution to depend upon the information regime. For example, on a good news day a buy order submitted at a lower price than the bid is less likely to be executed than on a no news day, because the preponderance of informed buy market orders will eventually cause upwards price movements. The same mechanism works for sell orders on a bad news day, so the parameters $\nu_I$ and $\nu_N$ represent the likelihood of limit order execution *relative* to a no news day.

---

[22]. Easley and O'Hara (1991) distinguish between two types of stop orders: A *stop-loss order* is an order that specifies a given quantity of stocks to be sold at a price that must be lower than the last trade price. A *stop-buy order* specifies a price at which the stock is bought, which must be higher than the last trade price. Orders to sell above the current bid or to buy below the current ask are known as *limit sell* respectively *limit buy orders*, see Sect. 2.2.2 for a description of different types of stop orders.

[23] Actually, they compare a market design, in which only the market maker has information on the order type to a market with a public limit order book, i.e. in which all traders observe the order type.

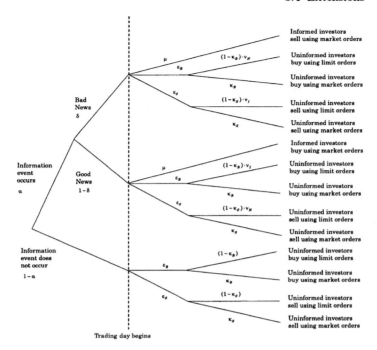

**Fig. 3.6.** The Easley, O'Hara and Saar (2001) model.

The model developed in Brown, Thomson, and Walsh (1999) provides a richer structure of the trading process by allowing for order size effects and no-trade events as well as for an informational role of the type of order. The introduction of the possibility of informed limit order submission is motivated by a different market design considered in their study, which is characterized by the absence of a single delegated market maker. Their model is intended to describe an electronic open limit order book instead, in which orders are automatically matched and executed by an automated trading system, so submitting limit orders is the natural way to generate trades. However, since all market participants may see the order book, it is possible to submit a limit order, that will match an already existing order in the opposite direction immediately. Trades generated in this way may be interpreted as market orders. Limit order submission by informed traders may be rational in this environment, if they want to avoid revelation of their trading motives by hiding in the noise of uninformed limit orders. Note, that their model is set up in terms of *order events* rather than trade events. The important difference is, that some of the orders submitted to the market may in fact never be executed.[24]

---

[24] In the Easley, O'Hara, and Saar (2001) model only limit orders that lead to trades are being considered.

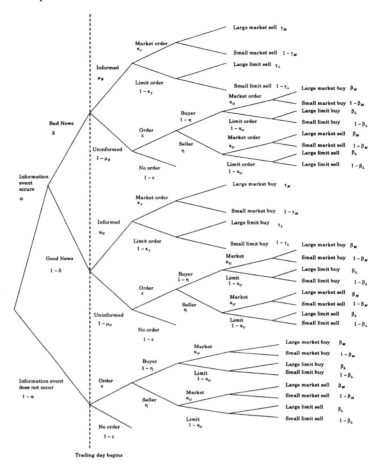

**Fig. 3.7.** The Brown, Thomson and Walsh (1999) model.

The information process at the beginning of the trading day is left unaltered, but the order submission process differs substantially from the models we considered before. First, the probabilities of choosing an informed trader is allowed to differ between good ($\mu_G$) and bad news days ($\mu_B$). Furthermore, the choice between market and limit orders is separately modelled for informed and uninformed traders. Informed traders choose market orders with probability $\kappa_I$, and decide to trade the large quantity with probability $\gamma_M$. When they choose to submit a limit order, the probability of trading the large quantity is given by $\gamma_L$. Note, that although both decisions, on the order form and on the quantity are independent of the type of information event, the observed fraction of, for example, large market sells submissions by informed traders may differ between good and bad news days, because of the different arrival rates of informed traders.

If an uninformed trader is chosen, he will decide to trade with probability $\varepsilon$ and choose to submit a sell order with probability $\eta$. The probability of submitting a market order is given by $\kappa_U$, and the large quantity is chosen with probability $\beta_M$ in this case, and with probability $\beta_L$ in case of limit order submission. The structure of this model variant is summarized by the tree diagram in Fig. 3.7.

Although the models by Easley, O'Hara, and Saar (2001) and Brown, Thomson, and Walsh (1999) allow for different effects of limit and market orders, they are not designed to be used for tests of the informational content of different order forms in the first place, but rather demonstrate the ease with which the sequential trade approach can be adapted to different study goals and market designs. Thus, the possibility to distinguish between limit and market orders adds to the inherent empirical content in coping with the basic premise of microstructure theory, namely to explore the extent to which the trading mechanism affects the price discovery process.

### 3.4.3 Models for Dually Listed Assets

The presence of informed trading poses a difficult problem for regulators concerned with the design of securities markets: If the market design is such, that it fully reveals all informed trades, then the price discovery process would be negatively affected, because incentives for costly information gathering activities would be diminished. But how can markets be efficient, if no one gathers information?[25] On the other hand, too much informational asymmetry will eventually destroy the market as market makers will set higher spreads to protect themselves from losing to informed traders and investors may be expected to switch to markets where trading costs are lower, even if they do not trade for speculative reasons in the first place.

Therefore, differences in market design between existing stock exchanges may be expected to affect the information content of the trading process. An important feature of the market design in this context is the degree of trader *anonymity*: On markets that require physical presence of traders and where the complete order flow goes through the hands of a delegated market maker, the possibility for informed traders to remain anonymous will be diminished, since repeated face-to-face interaction with market makers may enable them to identify informed traders ex post and to sanction[26] them in the future, if they refuse to reveal their information. On the other hand, in an electronic order

---

[25] This is the paradoxical implication of Grossman and Stiglitz (1980).

[26] In practice there are several ways, in which market makers may impose sanctions on traders, that they suspect of being informed: They may refuse to provide them better prices for their trades, e.g. by allowing them to execute an order inside the prevailing spread, refuse to fill orders above the quoted depth or be unwilling to help to execute a large order, see Heidle and Huang (1999).

driven market environment, individual traders typically cannot be identified by other market participants and so cannot be sanctioned. It is therefore plausible to assume, that markets with a higher degree of anonymity are more attractive for informed traders.

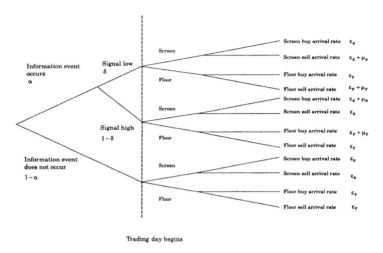

Trading day begins

**Fig. 3.8.** The Grammig, Schiereck and Theissen (2001) model.

In Grammig, Schiereck, and Theissen (2001) this issue is tackled in a sequential trade model, that allows for parallel trading on a *floor trading system* requiring physical presence of all traders and a *screen trading system*, that enables traders to remain anonymous. The model is a straightforward extension of the EKOP model, that additionally considers the decision on which market to trade. Both, informed and uninformed traders are assumed to trade on both markets. The arrival rates of uninformed floor (screen) traders are given by $\varepsilon_F$ ($\varepsilon_S$), while informed floor (screen) traders have arrival rates equal to $\mu_F$ ($\mu_S$). The structure of this model is summarized by the tree diagram in Fig. 3.8. The role of anonymity can be addressed by comparing the arrival rates of informed traders on both markets: If $\mu_S > \mu_F$, informed traders prefer the anonymous screen trading system to floor trading. In addition, the consequences of this preference for the quality of the market may be assessed, by computing the probability of informed trading for each market separately in analogy to (3.5) and use them for comparisons across markets.

A slightly different generalization of the basic model is used by Easley, Kiefer, and O'Hara (1996) to shed some light on the on-going debate about the practice of dealers and trading locals of alternative trading venues to pay brokers for redirecting order flow to their exchange. Typically, such payment for order flow agreements limit the size of the transactions they will accept and also require a minimum order flow that has to be redirected.[27] This practice is highly controversial. While the existence of different trading venues for the same asset may enhance competition between markets in general and thus lead to lower transaction costs by abolishing the capability to pocket monopoly profits in a single trading venue environment, competition also leads to a reduction of liquidity available in any of the venues, thus complicating the provision of a stable and orderly price process in practice.

Furthermore, the selective nature of the competition, when purchased order flow agreements are taken into account, may undermine not only the viability of markets not engaging in purchasing order flow but affect the trading process itself as well. One way to tackle this issue, is to note the adverse selection issue raised by this practice: By restricting the types of orders they will accept, some trading venues may try to attract only profitable liquidity traders, leaving larger, potentially more information based orders to other venues, thus exercising a practice known as *cream-skimming*. The structure of the model variant of Easley, Kiefer, and O'Hara (1996), displayed in Fig. 3.9, is designed to be used for tests of the extent to which purchased order flow may lead to cream-skimming in practice.

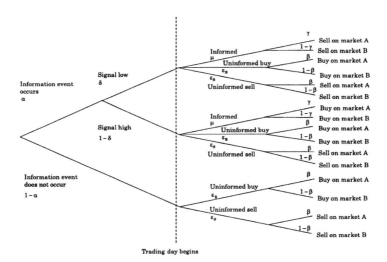

**Fig. 3.9.** The Easley, Kiefer and O'Hara (1996) model.

---

[27] Easley, Kiefer, and O'Hara (1996) report that in practice a significant fraction of about 35% of the small trades between 100 and 2099 shares have been redirected from the NYSE to alternative trading venues such as regional or electronic stock exchanges. Furthermore, these trades are purchased at 1 to 2 cents per share and minimum order flow specifications amount to at least 100,000 shares per month.

In this framework, informed trades arrive with rate equal to $\mu$, if an information event occurs, and are executed on trading venue $A$, that does not pay for order flow, with probability $\gamma$ and on the alternative trading venue $B$, that is purchasing order flow with probability $1 - \gamma$. The choice on which trading venue the order is executed is made by brokers, who are commissioned to execute trades, and is thus exogenous with respect to the choice set of investors. Uninformed buy (sell) orders arrive with $\varepsilon_B$ ($\varepsilon_S$) and are executed on market $A$ with probability $\beta$ and redirected to market $B$ with probability $1 - \beta$. On each trading venue there is a delegated market maker, who watches trades in both markets. Since market makers have common information, they will update their beliefs on the information regime by taking account of the order flow on both markets. Both market makers use Bayesian updating rules to set their bid and ask prices analogous to those derived for the basic model.

The key parameter that determines the information content of trades on either market in this setting is $\beta/\gamma$ the ratio of uninformed to informed trading propensities in market $A$. It can be shown that, in absence of any other factor that influences price setting behavior, the zero profit spread on market $A$ is decreasing with $\beta/\gamma$, while it is increasing on market $B$. The same holds true for the probability of informed trading, which can be shown to be equal to

$$PIT_A(\tau) = \frac{\mu \cdot (1 - p_n(\tau))}{\mu \cdot (1 - p_n(\tau)) + (\varepsilon_B + \varepsilon_S) \cdot \frac{\beta}{\gamma}},$$

on market $A$ and

$$PIT_B(\tau) = \frac{\mu \cdot (1 - p_n(\tau))}{\mu \cdot (1 - p_n(\tau)) + (\varepsilon_B + \varepsilon_S) \cdot \frac{1-\beta}{1-\gamma}},$$

on market $B$. When $\beta/\gamma = 1$ the information content of the order flow will be the same on both markets, and accordingly prices should be the same, while $\beta/\gamma < 1$ should lead to lower prices on market $B$. If however the prices on market $B$ are simply set equal to those on market $A$, then the market maker will earn positive expected profits and this will in turn allow him to purchase order flow. Thus, the selective nature of the composition of the order flow that is redirected to market $B$ is the critical factor, that makes payments for order flow agreements profitable.

Easley, O'Hara, and Srinivas (1998) derive a model for parallel trading in stock and option markets. The key question here is, whether derivative markets are used only for risk hedging purposes, or whether they also constitute a venue for informed traders. In their model of parallel trading, each market has his own market maker, who watches trades on both markets and thus learns from the combined order flow. Their behavior is otherwise standard, i.e. they act competitive, are risk neutral, and learn from the order flow according to the Bayesian learning model. On the option market there are two types of contracts available, *put* and *call* options, both having the same maturity, controlling the same number of shares $\varpi$, and both can be exercised only at the maturity date, i.e. they are European options.

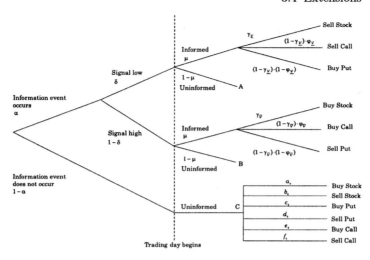

**Fig. 3.10.** The Easley, O'Hara and Srinivas (1998) model.

Informed traders can make profits either on the stock or the option market. On a bad news day informed traders know that the value of the stock at the maturity date of the option contract is $V = \underline{V}$. Their profit $\Pi$ over the holding period is equal to[28]

$$\underline{\Pi} = \begin{cases} b_S - \underline{V} & \text{if they sell the stock,} \\ -a_P + \varpi \cdot (X_P - \underline{V}) & \text{if they buy a put,} \\ b_C & \text{if they sell a call,} \end{cases}$$

where $b_S$ denotes the bid price for the stock, $X_P \in [\underline{V}, \overline{V}]$ denotes the exercise price of the put option, $a_P$ denotes the ask price for the put option and $b_C$ denotes the bid price of the call option. On a good news day profits are given by

$$\overline{\Pi} = \begin{cases} -a_S + \overline{V} & \text{if they buy the stock,} \\ b_P & \text{if they sell a put,} \\ -a_C + \varpi \cdot (\overline{V} - X_C) & \text{if they buy a call,} \end{cases}$$

where $a_S$ denotes the ask price for the stock, $X_C \in [\underline{V}, \overline{V}]$ denotes the exercise price of the call option, $b_P$ denotes the bid price for the put option and $a_C$ denotes the ask price of the call option. Each informed trader will choose to trade on the market, where his profits are maximized, given their information. The fraction of informed traders that will decide to trade on the stock market on a bad news day is denoted by $\gamma_{\underline{V}}$, and by $\gamma_{\overline{V}}$ on a good news day. The fraction of the informed that trade on the option market selling a call on a

---

[28] For simplicity we assume, that all stock transactions are for exactly one unit of the share.

bad news day is given by $\varphi_{\underline{V}}$ and $(1 - \varphi_{\underline{V}})$ is the fraction of informed traders buying a put. On a good news day these fractions are equal to $\varphi_{\overline{V}}$ and $1 - \varphi_{\overline{V}}$.

The behavior of the uninformed traders is allowed to be history dependent. Consider the branch of the tree diagram depicted in Fig. 3.10 at node C. Uninformed traders buy the stock with probability $a_\tau$, sell the stock with probability $b_\tau$, buy a put with probability $c_\tau$, and sell it with probability $d_\tau$. They will buy a call with probability $e_\tau$, and sell it with probability $f_\tau$. These probabilities are assumed to follow a stochastic process, that is independent of the information process, so the branches that follow the nodes A and B are identical to the branch at node C, and have been omitted from Fig. 3.10 for reasons of lucidity. Apart from the restrictions, that uninformed trading probabilities are measurable functions of the history, positive valued and that they sum to one, no further assumptions on uninformed behavior are being made.

It can be shown, that the bid and ask prices set by each market maker are functions of the overall probability of informed trading $\alpha \cdot \mu$ and the fraction of informed traders believed to be in each market $\gamma_{\underline{V}}$, $\gamma_{\overline{V}}$, $\varphi_{\underline{V}}$ and $\varphi_{\overline{V}}$. Furthermore, given the prices set by market makers, the informed will choose to trade where their profits are being maximized. This leads to the following condition for informed traders to use options

$$\varpi \cdot (X_P - \underline{V}) \text{ or } \varpi \cdot (\overline{V} - X_C) > \frac{(1 - \alpha \cdot \mu) \cdot b_\tau \cdot (\overline{V} - \underline{V})}{\alpha \cdot \delta \cdot \mu + b_\tau \cdot (1 - \alpha \cdot \mu)}.$$

There are two possible equilibria in this model.[29] In a *separating* equilibrium, all informed traders will choose to trade on the stock market, while in a *pooling* equilibrium, they will trade on both markets. A pooling equilibrium will be attained, when (a) the leverage effect of the options is large (i.e. if the number of shares controlled by the option $\varpi$ is relatively large) (b) the liquidity on the stock market is small (i.e. the overall fraction of uninformed traders choosing to trade the stock is relatively small) or (c) the number of informed traders is large. Otherwise informed traders will only be present at the stock market.

The Easley, O'Hara, and Srinivas (1998) approach was extended in a recent paper by Jern (2000), in which state dependent behavior of uninformed traders was accounted for by allowing their trading probabilities to depend on either the type of previous trades (positive or negative trade), the market on which the trade was conducted (stock or option market) or both. A positive (negative) trade in this context refers to any trade that would produce a positive expected gain in the presence of good (bad) news, i.e. informed traders would either engage in a buy (sell) on the stock market, buy (sell) a call, or sell (buy) a put on the option market. The main goal of his study is to asses the determinants of feedback between the two markets, which is related to the nature of state dependency of uninformed traders.

---

[29] This is reminiscent to the model variant with two different trading sizes derived in Easley and O'Hara (1987).

## 3.5 Estimation of Structural Models

### 3.5.1 Estimation of the Basic Model Using Information on Buys and Sells

Sequential trade models lend themselves directly to empirical applications. Difficulties in the estimation of structural parameters in these models arise because neither the occurrence of information events nor the associated arrival of informed and uninformed traders is directly observable. What is observable from the usual information contained in a high frequency data set instead is the sequence of trading events, such as the incoming buy and sell orders. We will describe the estimation techniques for the basic sequential trade model. Assuming we have a data set that contains daily counts on the number of sells $S_d$ and buys $B_d$ for a total of $D$ trading days, the parameter vector $\theta = (\alpha, \delta, \varepsilon, \mu)'$ can be estimated and inferences on the relevance of informed trading can be drawn employing standard maximum likelihood techniques.

The likelihood function can be derived as follows. If the trading process follows a Poisson process with arrival rates that depend on the type of the trading day, as is assumed in the EKOP framework, it is possible to recover information on the event type and the associated arrival rate from the observed intensity of the trading process on different days in the sample. For example on a bad news day one would expect to observe more sells than buys, while on a good news day the opposite should hold true. On a no news day, we would expect to have fewer trades than usual with the number of sells and buys being approximately balanced. Thus, the unconditional distribution of the number of daily trades will be a mixture of three different Poisson distributions with parameters depending on the type of trading day. This information can be exploited to draw inference on the structural parameters $\theta$.

On a bad news day, sell orders arrive with rate $(\mu + \varepsilon)$ while buy orders arrive with rate $\varepsilon$. Since we assumed that the order arrivals of informed and uninformed traders are independent of each other, the density of a sequence of orders, containing $B_d$ buys and $S_d$ sells, given that the current trading day $d$ is a bad news day with a trading interval of length $\Delta\tau$, is given by

$$f(B_d \cap S_d \mid \text{bad news}; \theta) = \exp\left[-\varepsilon \cdot \Delta\tau\right] \cdot \frac{[\varepsilon \cdot \Delta\tau]^{B_d}}{B_d!} \tag{3.8}$$

$$\cdot \exp\left[-(\mu + \varepsilon) \cdot \Delta\tau\right] \cdot \frac{[(\mu + \varepsilon) \cdot \Delta\tau]^{S_d}}{S_d!}.$$

The conditional densities of the trading process on a good news day respectively no news day can be derived in the same manner. These are equal to

$$f(B_d \cap S_d \mid \text{good news}; \theta) = \exp\left[-(\mu + \varepsilon) \cdot \Delta\tau\right] \cdot \frac{[(\mu + \varepsilon) \cdot \Delta\tau]^{B_d}}{B_d!} \tag{3.9}$$

$$\cdot \exp\left[-\varepsilon \cdot \Delta\tau\right] \cdot \frac{[\varepsilon \cdot \Delta\tau]^{S_d}}{S_d!},$$

$$f\left(B_d \cap S_d \mid \text{no news}; \theta\right) = \exp\left[-\varepsilon \cdot \Delta \tau\right] \cdot \frac{\left[\varepsilon \cdot \Delta \tau\right]^{B_d}}{B_d!} \tag{3.10}$$

$$\cdot \exp\left[-\varepsilon \cdot \Delta \tau\right] \cdot \frac{\left[\varepsilon \cdot \Delta \tau\right]^{S_d}}{S_d!}.$$

In practice the type of day is unknown to the researcher. In order to derive the unconditional probability of observing a total of $B_d$ buys and $S_d$ sells on an arbitrary type of day, we exploit the independence assumption and find

$$\begin{aligned}
f_d\left(B_d \cap S_d; \theta\right) = {} & \alpha\delta \cdot f\left(B_d \cap S_d \mid \text{bad news}; \theta\right) \tag{3.11} \\
& + \alpha\left(1 - \delta\right) \cdot f\left(B_d \cap S_d \mid \text{good news}; \theta\right) \\
& + \left(1 - \alpha\right) \cdot f\left(B_d \cap S_d \mid \text{no news}; \theta\right).
\end{aligned}$$

Note that the preceding equation constitutes an example of a mixture of three independent Poisson distributions, each describing the aggregate behavior of the market conditional on the unobservable type of information event on a particular day. Given a data set with a total number of $D$ trading days in the sample, the log likelihood function is given by the log of the product of the unconditional densities for each separate day

$$\ln \mathcal{L}\left(\theta\right) = \sum_{d=1}^{D} \ln f_d\left(B_d \cap S_d; \theta\right). \tag{3.12}$$

In addition, the following restrictions on the parameter vector $\theta$ have to be imposed: $\alpha, \delta \in [0, 1]$ and $\mu, \varepsilon \in [0, \infty)$. Since the derivatives of the log likelihood are non-linear, some iterative maximization procedure has to be employed in order to find the estimates $\hat{\theta}$. We will introduce an alternative estimation technique for a generalized variant of the EKOP-model based on the EM-algorithm in Sect. 4.1.2.

### 3.5.2 Estimation of the Basic Model Using Information on Trades

In the preceding Section we derived a ML-estimator for the parameters of the EKOP model based on a sample of observations on two time series, the number of daily buys $B_d$ and the number of daily sells $S_d$. In many circumstances, information on the *trade direction*, i.e. whether a particular transaction was initiated by a buyer or a seller, is not contained in the data set available for estimation, but has to be recovered by applying an appropriate classification algorithm.[30] Since any algorithm that has been proposed to classify trades bears the danger of misclassifying trades, the resulting parameter estimates may be biased.[31]

---

[30] See Sect. 5.2 for a discussion.

[31] Recently Grammig and Theissen (2002) showed by means of a Monte Carlo study that misclassification leads to biased estimates of the trade arrival rates $\mu$ and

An attractive alternative to the use of classification algorithms, is to estimate the parameters of the EKOP model from the pure trading process, which is directly observable. The random variable $A_d \equiv B_d + S_d$ equals the total number of daily trades. Note that from the countable additivity theorem[32] under the bad news regime $A_d$ has conditional density equal to

$$f\left(A_d \mid \text{bad news}; \theta\right) = \exp\left[-\left(2 \cdot \varepsilon + \mu\right) \cdot \Delta\tau\right] \cdot \frac{\left[\left(2 \cdot \varepsilon + \mu\right) \cdot \Delta\tau\right]^{A_d}}{A_d!}, \quad (3.13)$$

while the corresponding densities under the good and no news regimes are given by

$$f\left(A_d \mid \text{good news}; \theta\right) = \exp\left[-\left(2 \cdot \varepsilon + \mu\right) \cdot \Delta\tau\right] \cdot \frac{\left[\left(2 \cdot \varepsilon + \mu\right) \cdot \Delta\tau\right]^{A_d}}{A_d!}, \quad (3.14)$$

and

$$f\left(A_d \mid \text{no news}; \theta\right) = \exp\left[-\left(2 \cdot \varepsilon\right) \cdot \Delta\tau\right] \cdot \frac{\left[\left(2 \cdot \varepsilon\right) \cdot \Delta\tau\right]^{A_d}}{A_d!}. \quad (3.15)$$

Thus, the unconditional density function of the trades is given by

$$
\begin{aligned}
f_d\left(A_d; \theta\right) = {} & \alpha \cdot \delta \cdot f\left(A_d \mid \text{bad news}; \theta\right) \quad (3.16) \\
& + \alpha \cdot (1 - \delta) \cdot f\left(A_d \mid \text{good news}; \theta\right) \\
& + (1 - \alpha) \cdot f\left(A_d \mid \text{no news}; \theta\right) \\
= {} & \alpha \cdot f_j\left(A_d \mid \text{bad news}; \theta\right) + (1 - \alpha) \cdot f_j\left(A_d \mid \text{no news}; \theta\right),
\end{aligned}
$$

so the corresponding trading process $A_d$ will be governed by a mixture of two independent Poisson distributions with mixing probability equal to the unconditional probability of a news event $\alpha$. The parameter $\delta$, the unconditional probability of observing a bad news event cannot be identified from the sequence of trades. This means we can estimate whether *any* news event had occurred or not, but we can not determine whether it was good or bad news. The corresponding likelihood function may be derived from (3.16). If one is only concerned with estimating the overall share of informed trading *PIT*, estimation based on the time series of trades $A_d$ is the preferable alternative compared to the original EKOP method based on the bivariate process

---

$\varepsilon$, as well as biased estimates of the probability of informed trading *PIT* in the EKOP model. Estimates of the regime probabilities $\alpha$ and $\delta$ tend to be unaffected by the bias.

[32] The *countable additivity theorem* states that, if $y_1, y_2, \ldots$ are independent random variables, each having a Poisson distribution with mean $\lambda_k$, and if $\lambda_y = \sum_{k=1}^{\infty} \lambda_k$ is a convergent series, then $y = \sum_{k=1}^{\infty} y_k$ also converges with probability 1 and $y$ is a proper random variable having a Poisson distribution with mean equal to $\lambda_y$. See Kingman (1993), pp. 5-6 for a proof.

for buys and sells. The advantage comes in terms of a reduction of computational effort and an avoidance of biased estimates resulting from possible misclassification of trades.[33]

The parameter $\delta$ can be identified unambiguously by relaxing some of the assumptions implicit in the EKOP framework, e.g. by letting informed buyers and sellers have different arrival rates. This might be justified by short selling constraints that some of the informed traders may face in practice. If informed traders are not able to circumvent these constraints then it is reasonable to assume that the sell arrival rate $\mu_S$ is lower than the buy arrival rate $\mu_B$ and thus the parameter $\delta$ is identified by $\mu_B > \mu_S$.[34] In this generalized form of the EKOP model the unconditional density function of the trades is given by the three component mixture

$$
\begin{aligned}
f_d\left(A_d; \theta\right) = \quad & \alpha \cdot \delta \cdot \exp\left[-\left(2 \cdot \varepsilon + \mu_S\right) \cdot \Delta\tau\right] \cdot \frac{\left[\left(2 \cdot \varepsilon + \mu_S\right) \cdot \Delta\tau\right]^{A_d}}{A_d!} \quad (3.17) \\
+ & \alpha \cdot (1 - \delta) \cdot \exp\left[-\left(2 \cdot \varepsilon + \mu_B\right) \cdot \Delta\tau\right] \cdot \frac{\left[\left(2 \cdot \varepsilon + \mu_B\right) \cdot \Delta\tau\right]^{A_d}}{A_d!} \\
+ & (1 - \alpha) \cdot \exp\left[-\left(2 \cdot \varepsilon\right) \cdot \Delta\tau\right] \cdot \frac{\left[\left(2 \cdot \varepsilon\right) \cdot \Delta\tau\right]^{A_d}}{A_d!}.
\end{aligned}
$$

An extension of this approach is used by Easley, Engle, O'Hara, and Wu (2002), who incorporate time varying arrival rates of informed and uninformed traders into the basic sequential trade model. Their model is based on the bivariate stochastic process of the *daily absolute trade imbalance* $K_d \equiv |\, S_d - B_d \,|$, and the *daily number of balanced trades* $Q_d \equiv A_d - K_d$, whose properties can easily be derived from the basic model.[35] First, note that the unconditional (regime unspecific) mean of the daily number of trades $A_d$ in the original EKOP framework is given by

$$
\begin{aligned}
E(A_d) = \; & \alpha \cdot \delta \cdot E(A_d \mid \text{bad news}) + \alpha \cdot (1 - \delta) \cdot E(A_d \mid \text{good news}) \\
& + (1 - \alpha) \cdot E(A_d \mid \text{no news}) \\
= \; & \alpha \cdot \delta \cdot (2 \cdot \varepsilon + \mu) + \alpha \cdot (1 - \delta) \cdot (2 \cdot \varepsilon + \mu) + (1 - \alpha) \cdot (2 \cdot \varepsilon) \\
= \; & \alpha \cdot \mu + 2 \cdot \varepsilon.
\end{aligned}
$$

Second, similar calculations reveal that the unconditional mean of the *trade imbalance* $S_d - B_d$ is given by $E(S_d - B_d) = \alpha \cdot \mu \cdot (2\delta - 1)$, while the expectation of the absolute trade imbalance $K_d$ is approximately equal to

---

[33] Easley, Kiefer, and O'Hara (1997b) exploited this feature in a variant of the sequential trade model to test for time dependence of the information regime.

[34] This is motivated by the implications of Diamond and Verrecchia (1987), who showed that short selling constraints eliminate informative trades and reduce the adjustment speed of prices to private information when there is bad news.

[35] Their use of the trade imbalance rather than the actual number of buy or sell trades might be inspired by a recent empirical study by Chordia, Roll, and Subrahmanyam (2002) on the causes of order imbalances.

$$E(K_d) \approx \alpha \cdot \mu, \tag{3.18}$$

when $\mu$ is large. Therefore, the composite random variable $Q_d$, which can be interpreted as the daily number of balanced trades, has unconditional expected value equal to

$$E(Q_d) = 2 \cdot \varepsilon. \tag{3.19}$$

Note, that the bivariate stochastic process $Z_d = [K_d, Q_d]'$ contains all the essential information on the arrival rates of uninformed and informed traders. In the original EKOP framework, both variables are independently identically distributed (i.i.d.) and should be uncorrelated with each other. Easley, Engle, O'Hara, and Wu (2002) allow for both auto- and crosscorrelation between the elements of $Z_d$, by proposing a generalized autoregressive conditional mean specification that is analogous to the conditional mean specification in GARCH models[36]

$$\psi_d = \omega + \sum_{i=1}^{p} \Phi_i \cdot \psi_{d-i} + \sum_{j=0}^{q-1} \Gamma_j \cdot Z_{d-j}, \tag{3.20}$$

where the $(2 \times 1)$ vector $\psi_d$ denotes the conditional expectation of $Z_d$, $\omega$ is a $(2 \times 1)$ vector of constants, and $\Phi_i$ and $\Gamma_j$ are $(2 \times 2)$ matrices of coefficients. Easley, Engle, O'Hara, and Wu (2002) also propose an alternative autoregressive specification for the logarithm of $\psi_d$ that is analogous to the EGARCH model of Nelson (1991). Estimation of the bivariate vector process is straightforward, by simply assuming a conditional Poisson density for each of the two observable time series $S_d$ and $B_d$, and replacing the constant arrival rates $\mu$ and $\varepsilon$ in (3.11) by time varying arrival rates $\mu_d \equiv \alpha^{-1} \cdot \psi_{d1}$, and $\varepsilon_d \equiv 0.5 \cdot \psi_{d2}$, where $\psi_{d1}$ denotes the conditional expectation of $K_d$ and $\psi_{d2}$ the conditional expectation of $Q_d$. Note that this parameterization implies, that the probability of a news event $\alpha$ is time invariant. This assumption is needed in order to achieve identification.

### 3.5.3 Estimation of Related Models

The conditional independence assumption that has been exploited to derive the joint density of the trading events conditional on the information regime[37] is the key step in the derivation of the likelihood function in model variants that distinguish between more than two trading events or have more than three information regimes.[38] In some of the variants of the sequential trade

---

[36] See Bollerslev (1986).

[37] See (3.8) through (3.10).

[38] Thus, the modifications needed to estimate the structural parameters in the models by Easley, Kiefer, and O'Hara (1996), Easley, O'Hara, and Saar (2001), Weston (2001), and Grammig, Schiereck, and Theissen (2001) are straightforward extensions of the basic approach.

model however either no explicit assumption on the conditional density of the trading events has been made or the models were formulated based on alternative specifications. For example, in Easley, Kiefer, and O'Hara (1997b) estimation of the parameters in the model variant without trade size effects was based on the assumption of a multinomial conditional distribution of the three trading events, buys, sells and no-trade intervals. The probability of observing $B_d$ buys, $S_d$ sells and $NT_d$ no-trade events on a bad news day is thus given by

$$
f(B_d \cap S_d \cap NT_d \mid \text{bad news}; \theta) = \Xi_d \cdot [(1-\mu) \cdot 0.5 \cdot \varepsilon]^{B_d} \qquad (3.21)
$$
$$
\cdot [\mu + (1-\mu) \cdot 0.5 \cdot \varepsilon]^{S_d}
$$
$$
\cdot [(1-\mu) \cdot (1-\varepsilon)]^{NT_d},
$$

where the combinatorial factor $\Xi_d$ is given by

$$
\Xi_d \equiv \left[ \frac{(B_d + S_d + NT_d)!}{B_d! \cdot S_d! \cdot NT_d!} \right],
$$

and the parameter vector is equal to $\theta = (\alpha, \delta, \mu, \varepsilon)'$.[39] The corresponding probabilities for the good news, respectively no news days are given by

$$
f(B_d \cap S_d \cap NT_d \mid \text{good news}; \theta) = \Xi_d \cdot [\mu + (1-\mu) \cdot 0.5 \cdot \varepsilon]^{B_d} \qquad (3.22)
$$
$$
\cdot [(1-\mu) \cdot 0.5 \cdot \varepsilon]^{S_d}
$$
$$
\cdot [(1-\mu) \cdot (1-\varepsilon)]^{NT_d},
$$
$$
f(B_d \cap S_d \cap NT_d \mid \text{no news}; \theta) = \Xi_d \cdot [0.5 \cdot \varepsilon]^{B_d + S_d} \cdot (1-\varepsilon)^{NT_d}. \quad (3.23)
$$

The log-likelihood function is derived from the unconditional joint density function of the three trading events, which again is a mixture of the three conditional densities, weighted by the probabilities of observing the information regimes

$$
f_d(B_d \cap S_d \cap NT_d; \theta) = \alpha \cdot \delta \cdot f(B_d \cap S_d \cap NT_d \mid \text{bad news}; \theta) \qquad (3.24)
$$
$$
+ \alpha \cdot (1-\delta) \cdot f(B_d \cap S_d \cap NT_d \mid \text{good news}; \theta)
$$
$$
+ (1-\alpha) \cdot f(B_d \cap S_d \cap NT_d \mid \text{no news}; \theta).
$$

The log-likelihood function for a sample of $D$ trading days may be computed in the usual manner. The likelihood function for the second model variant of Easley, Kiefer, and O'Hara (1997b), in which trade size effects are allowed, as well as for the model of Brown, Thomson, and Walsh (1999) may be derived in a completely analogous way.

The model of Easley, Kiefer, and O'Hara (1997a) contains an additional feature, a time dependent parameterization of the probabilities of trading ($\varepsilon_\tau$)

---

[39] Note, that the probabilities may be read off from the branches of the tree diagram given in Fig. 3.2.

and selling $(\eta_\tau)$ of the uninformed traders. In this model, there are five different trading events [no-trade (NT), large sells (LS), small sells (SS), large buys (LB) and small buys (SB)] and eleven parameters to be estimated in total. The corresponding likelihood function is therefore quite complex, but the basic principle of derivation remains unchanged. The conditional density of the observed trading events will in general depend on the sequence of trades within each day. For example, on a bad news day the probabilities of observing a no-trade interval at any time $\tau$ during the trading day will depend on the type of trading event observed at time $\tau - 1$. There are five possible transitions between successive trading events and the corresponding probabilities are given by

$$f(NT_\tau \mid NT_{\tau-1}, \text{bad news}; \theta) = (1 - \mu) \cdot [1 - \varepsilon(NT)],$$
$$f(NT_\tau \mid LS_{\tau-1}, \text{bad news}; \theta) = \mu \cdot \gamma + (1 - \mu) \cdot [1 - \varepsilon(S)],$$
$$f(NT_\tau \mid SS_{\tau-1}, \text{bad news}; \theta) = \mu \cdot (1 - \gamma) + (1 - \mu) \cdot [1 - \varepsilon(S)],$$
$$f(NT_\tau \mid LB_{\tau-1}, \text{bad news}; \theta) = (1 - \mu) \cdot [1 - \varepsilon(B)],$$
$$f(NT_\tau \mid SB_{\tau-1}, \text{bad news}; \theta) = (1 - \mu) \cdot [1 - \varepsilon(B)],$$

with parameter vector $\theta$ given by

$$\theta = [\alpha, \delta, \mu, \gamma, \beta, \varepsilon(NT), \varepsilon(B), \varepsilon(S), \eta(NT), \eta(B), \eta(S)]'.$$

The probabilities for the remaining four trading events and for the good and no news days may be derived in the same fashion, i.e. by following the branches of the tree diagram in Fig. 3.4, replacing $\varepsilon_\tau$ and $\eta_\tau$ by their appropriate values and multiplying the resulting probabilities along each branch. The contribution to the likelihood function of any specific trading day is then obtained by multiplying the conditional densities of the observed sequences of trading events, thereby taking the first trade on each day as given, weighting them with the corresponding probabilities of the information events and summing these weighted likelihood contributions in order to obtain the unconditional density. The log-likelihood for the total sample period, consisting of $D$ trading days is then found by summing the logarithms of the daily likelihood contributions over the sample period.

Although neither Jern (2000) nor Lei and Wu (2001) control for different trade size, their estimation methods are very similar in spirit to Easley, Kiefer, and O'Hara (1997a). The likelihood function for the parallel trading model with time dependent trading probabilities for uninformed traders considered by Jern (2000) can be derived in basically the same manner as stated above. The model variant proposed by Lei and Wu (2001) allows both uninformed and informed arrival rates to depend on observable characteristics of the trading process, e.g the historical transaction volume, or the return on the S&P 500 composite stock market index and additionally employs a distinct distributional assumption.[40] It therefore requires some nontrivial modifications of the approach discussed above.

While the distributional assumption requires only a minor refinement of the likelihood function, the inclusion of explanatory variables is achieved via the introduction of an unobservable two state Markov chain, that governs the evolution of uninformed trade arrival rates. Specifically, Lei and Wu (2001) allow uninformed trade arrival rates to be either at a high level $\varepsilon_\tau^{(h)}$ or at a low level $\varepsilon_\tau^{(l)}$, with transition probabilities between the high and low level states collected in a $(2 \times 2)$ transition matrix $P$ given by

$$
P = \begin{bmatrix} p_\tau^{(ll)} & 1 - p_\tau^{(hh)} \\ 1 - p_\tau^{(ll)} & p_\tau^{(hh)} \end{bmatrix},
$$

where $p_\tau^{(ll)} \equiv p(\varepsilon_\tau^{(l)} \mid \varepsilon_{\tau-1}^{(l)})$ denotes the probability of observing the low level trading activity state in two successive periods $\tau - 1$ and $\tau$ and $1 - p_\tau^{(ll)} \equiv p(\varepsilon_\tau^{(h)} \mid \varepsilon_{\tau-1}^{(l)})$ denotes the probability of observing a transition from the low activity level state in $\tau - 1$ to the high activity level state in $\tau$. The transition probabilities for the high activity level state $p_\tau^{(hh)}$ and $1 - p_\tau^{(hh)}$ are defined analogously. The likelihood of observing a transition between the two states depends on a set of exogenous variables collected in a column vector $x_\tau$, which is related to the transition probabilities via a logistic transformation, i.e.

$$
p_\tau^{(ll)} = \frac{\exp(x_\tau' \beta^{(l)})}{1 - \exp(x_\tau' \beta^{(l)})},
$$

$$
p_\tau^{(hh)} = \frac{\exp(x_\tau' \beta^{(h)})}{1 - \exp(x_\tau' \beta^{(h)})},
$$

where $\beta^{(l)}$ and $\beta^{(h)}$ are column vectors of state dependent regression parameters conformable with $x_\tau$. Time dependent arrival rates for informed traders can be incorporated into this framework in analogous manner. Estimation then proceeds by deriving the likelihood contributions for each trading day, conditional on being either in the low or the high activity level state and on the information regime, weighting them by the regime probabilities in order to obtain the unconditional likelihood contributions, and maximizing the corresponding log-likelihood function for a sample of $d$ trading days by numerical methods. Further details on this procedure may be found in Lei and Wu (2001).

## 3.6 Results of Previous Studies

The general modus operandi in many empirical studies is to estimate the structural model for a sample of stocks, conduct a few more or less demanding

---

[40] Their model is set up in continuous time, so they assume marginal Poisson distributions for the trading events.

specification tests, and, when the test results indicate no misspecification, to compute the probability of informed trading in analogy to (3.5), and to use these in turn for cross-sectional regressions, thus enabling comparisons of the information content of trading activity between groups of stocks. It turns out, that the basic EKOP-model has been the most frequently used basis in empirical studies for this task. A summary of the types of models, the estimation methods, the data sets used and the main results of these studies is given in Table 3.1.

Both, Easley, Kiefer, O'Hara, and Paperman (1996) and Grammig, Schiereck, and Theissen (1999) use estimates of the probability of informed trading to explain differences in the average level of the bid-ask spread in cross-sections of frequently and infrequently traded stocks. While the first study finds, that the probability of information based trading declines with the frequency of trading, the second study does not support this conclusion. However, this should not be interpreted as counterfactual evidence, since the sample of stocks used in Grammig, Schiereck, and Theissen (1999) are all contained in the German stock market index DAX, which includes the most frequently traded stocks only.

The basic model is also used by Easley, O'Hara, and Paperman (1998) to investigate the role of financial analysts, by Heidle and Huang (1999) to explore the effects on information based trading, when firms switch the stock exchange they are listed on, by Easley, Hvidkjaer, and O'Hara (2001) to compare the predictive power of information based trading on expected stock returns in an extended variant of the capital asset pricing model (CAPM), and by Beber and Caglio (2002) to investigate the role of information asymmetry on the choice of order submission strategies.

A variant of the basic model that allows for limit order submission by uninformed traders is used by Easley, O'Hara, and Saar (2001) to examine the effects of stock splits on the trading activity, by estimating their model separately for a time window before and after the split for each stock in their sample. The sample window splitting technique has also been used by Heidle and Huang (1999), Fu (2002) and Aktas, de Bodt, Declerck, and Van Oppens (2003). Fu (2002) examines informational aspects of equity carve-outs employing the basic model.[41] Using a sample of firms that undertake an equity carve-out and an industry and market value matched control sample, the study documents decreases of the average probability of a news event and the probability of informed trading after the equity carve-out, while firms in the control sample exhibit no systematic changes in these quantities. Aktas, de Bodt, Declerck, and Van Oppens (2003) examine the behavior of the probability of informed trading around public merger and acquisition announcements and find, contrary to their intuition, that it increases after the announcement date.

---

[41] An equity carve-out occurs when a company sells a portion of their equity in a subsidiary to the public.

The evidence on the information content of trade size appears to be mixed. While Easley, Kiefer, and O'Hara (1997b) find that trade size is uninformative, both Easley, Kiefer, and O'Hara (1997a) and Brown, Thomson, and Walsh (1999) find that trade size matters, although in different ways. The results of Easley, Kiefer, and O'Hara (1997a) indicate, that informed traders choose large trades with a higher probability than uninformed traders, whereas Brown, Thomson, and Walsh (1999) find on the contrary that informed traders choose small orders with higher probability than uninformed traders. However, both studies find evidence, that informed traders generally prefer small orders to large orders. Furthermore, all three studies agree that the occurrence of no-trade intervals is informative, thus underscoring the notion that no trading means no news.

With respect to the type of order submission chosen by traders, the findings of Brown, Thomson, and Walsh (1999) indicate, that informed traders appear to be undistinguishable from uninformed traders in their choice of limit and market orders, thus leading to the conclusion that knowledge of the limit order book does not lead to informational advantages per se. Beber and Caglio (2002) find on the contrary, that informed traders differ from uninformed traders in their order submission behavior. Their results are consistent with the notion that informed traders either try to hide their superior information or are constraint by short-selling restrictions since they appear to submit passive limit orders more likely than market orders. Note, that these findings may not be compared to the results of Easley, O'Hara, and Saar (2001), as their model variant excludes the possibility of informed limit order submission a priori.

There are several versions of the sequential trade model, that allow for a more complex behavior of uninformed traders. The studies of Easley, Kiefer, and O'Hara (1997a), Brown, Thomson, and Walsh (1999), Jern (2000), and Easley, Engle, O'Hara, and Wu (2002) find evidence, that uninformed trade arrival rates are indeed history dependent. In addition, the results of Lei and Wu (2001) suggest, that informed traders try to mimic the behavior of uninformed traders in an attempt to disguise their superior information.

An alternative explanation for observed serial dependence patterns in trade arrivals is given in the noise trading model of Weston (2001). His empirical results suggest, that at least part of the observed variation in trade arrival rates might not be caused by learning of uninformed, but by the presence of a third type of traders, noise traders, who submit additional orders on both sides of the market whenever information events occur. Given that the population of noise traders may be comprised of investors, who are likely to follow technical trading rules and are receptive for advise by stock analysts, these findings are consistent with Easley, O'Hara, and Paperman (1998). They report that stocks with a high degree of analyst coverage have higher arrival rates of both, informed and uninformed traders, which is somewhat surprising, since the behavior of informed traders would be expected to have no relation to any publicly available information including reports published by stock analysts. Their evidence might therefore indeed be caused by the presence of noise traders who were incorrectly classified either as informed or uninformed liquidity traders.

Easley, O'Hara, and Srinivas (1998) analyze simultaneous trading in stock and option markets. They do not estimate structural parameters, but rather test some implications of their sequential trade model by using tests for Granger causality between stock prices and option volume time series. They find evidence consistent with the notion that option markets are indeed a venue for informed traders.[42] In Jern (2000) the nature of the observed feedback effects between option and stock markets was more closely investigated by using an enhanced version of the Easley, O'Hara, and Srinivas (1998) model. He finds, that although informed traders typically submit orders on both markets, they prefer trading stocks to options. Feedback effects between markets are therefore not solely explained by asymmetric information among the population of investors, but appear to be driven by attempts of uninformed traders to learn from the observed order flow on both markets, and are also consistent with arbitrage and hedging considerations.

Easley, Kiefer, and O'Hara (1996) investigate the relation between trading on two regional stock exchanges and find evidence that the practice of purchasing order flows leads to cream-skimming, i.e. it attracts liquidity based trades in the first place. Grammig, Schiereck, and Theissen (2001) are concerned with simultaneous stock trading in an electronic and a floor trading environment. They focus on the relation between trader anonymity and information based trading and find that anonymity is associated with a higher share of information based trades. These results are in line with the findings of Heidle and Huang (1999), although they do not analyze parallel trading, but rather examine firms that switch between stock exchanges with different degrees of anonymity.

The consideration of parallel trading, limit order submission, trade size effects and history dependence of uninformed trading, may be viewed as attempts to incorporate elements of strategic behavior into the sequential trade approach. On the other hand, these models are still highly stylized in the sense that they abstract from many important features of real world trading environments. Also, the limitation to only two possible trade sizes and the definition of the no-trade event by some ad hoc definition of the pre-trade interval introduces some elements of arbitrariness into the framework. While this might be viewed as an inevitable simplification in order to keep the economic model tractable, their use in empirical studies might be more restrictive than they may seem at first glance.

Another critical point to be made, is that virtually all of the empirical studies discussed so far, estimate structural parameters based on time series

---

[42] In a related study using trade data on option and future contracts for the German stock index DAX, Schlag and Stoll (2001) find that informed traders prefer trading in future contracts over options.

of daily trading events.[43] If the data generating process of the observed time series is indeed a simple mixture of Poisson processes, this would be a perfectly legitimate way to proceed. However, there are several studies[44] that document, that tick by tick data sets obtained from several stock exchanges around the world exhibit characteristics, like serial correlation and intradaily seasonal patterns, that do not fit very well with the highly stylized description of the trading process implied by the sequential trade framework.

Hujer, Vuletić, and Kokot (2002) employ an alternative empirical approach which is tailor-made for estimation based on intradaily data sets. Their econometric model combines the autoregressive conditional duration (ACD) model introduced by Engle and Russell (1998), and the Markov regime switching model of Hamilton (1989). Introducing the Markov switching ACD model (MSACD) they test the implications of the EKOP model indirectly and find that the implied time varying regime probabilities derived from their empirical model for a time series of durations between successive trades have predictive power in explaining the inclination to buy. This corresponds to the evidence presented by Lei and Wu (2001) and Easley, Engle, O'Hara, and Wu (2002), who find, that arrival rates of informed and uninformed traders tend to be time varying. However, both studies use long series of daily trade event counts that span several years.

In the following parts of this study we will introduce an econometric framework, that will enable us to account for seasonal patterns as well as for serial correlation in several versions of the basic sequential trade model. Our approach will be most intimately related to the approach introduced by Hujer, Vuletić, and Kokot (2002), since we will employ a Markov switching approach with autoregressive dynamics. We will apply this framework to the estimation of a multivariate Poisson regression model. Thus, we will continue to aggregate the number of trading events, albeit over some finer time interval, than a whole trading day. Since our model is based on a multivariate approach, we can estimate empirical models that use information on the trade direction by employing time series of buy and sell transactions. The univariate MSACD model of Hujer, Vuletić, and Kokot (2002) on the other hand relies on using information on trades only. Therefore, our approach is much closer in spirit to the original sequential trade framework than the model of Hujer, Vuletić, and Kokot (2002) and is also able to explain well known empirical features of the trading process that have not been examined in other empirical studies so far.

---

[43] The only exception to this practice is the study by Brown, Thomson, and Walsh (1999), who divide the trading day into three parts in order to investigate, whether arrival rates change systematically during the day in a simplified version of their model.

[44] See Sect. 4.1.1 for details.

Table 3.1. Empirical research using sequential trade models

| Authors | Model-type | Estimation method | Data set | Main results |
|---|---|---|---|---|
| Easley, Kiefer, O'Hara, and Paperman (1996) | Basic STM. | MLE based on 2 different trading events, assuming conditional Poisson distributions for the trading events, given the information regime. | Daily trade event counts for 90 stocks from three different volume deciles (high, medium, and low volume) traded on the NYSE from the ISSM for the period October 1, 1990 until December 23, 1990 (60 trading days). | PIT is generally low for high volume stocks and virtually identical for medium and low volume stocks. The average bid-ask spread follows this pattern across volume deciles: It is low for high volume stocks, and about the same for medium and low volume stocks. |
| Easley, Kiefer, and O'Hara (1996) | Extension of basic STM with consideration of two parallel regional markets. | MLE based on 4 trading events; otherwise identical to Easley, Kiefer, O'Hara, and Paperman (1996). | Daily trade event counts for the 30 most actively traded stocks, that enlist on both the NYSE and CSE from the ISSM database for the period October 1 until December 23, 1990 (60 trading days). | PIT is larger on the NYSE than on CSE. |
| Easley, Kiefer, and O'Hara (1997b) | Two STM variants with explicit consideration of no-trade and trade intervals and trade size effects. | MLE based on 3 (model without trade size effects) and 5 (model with trade size effects) trading events, assuming conditionally multinomial distributions for the trading events, given the information regime. | Daily trade event counts for one stock traded on the NYSE (Ashland Oil) from the ISSM database for the period October 1 until December 22, 1990 (60 trading days). | A comparison of estimates for the STM variant with and without trade size effects yields no evidence in favour of the model with trade size effects. The trade size appears to be uninformative with respect to the information regime. |

AMEX = American stock exchange, ASX = Australian stock exchange, BODB = Berkley options database, CBOE = Chicago board options exchange, CSE = Cincinnati stock exchange, DAX = Deutscher Aktienindex (German stock market index), DTB = Deutsche Terminbörse (German options exchange), FSE = Frankfurt stock exchange, I/B/E/S = Institutional brokers estimation system, IBIS = Integriertes Börsenhandels- und Informationssystem (Integrated trading and informationsystem), ISSM = Institute for the study of security markets, MLE = Maximum likelihood estimation, MSACD = Markov switching autoregressive conditional duration, NASDAQ = National association of secruities dealers automated quotation, NYSE = New York stock exchange, PIT = Probability of informed trading, PNT = Probability of noise trading, SIRCA = Securities industry research centre of the Asia-Pacific, STM = Sequential trade model, TAQ = Trades and quotes, TORQ = Trades, orders, reports, and quotes.

**Table 3.1**: Empirical research using sequential trade models (cont.)

| Authors | Model-type | Estimation method | Data set | Main results |
|---|---|---|---|---|
| Easley, Kiefer, and O'Hara (1997a) | STM with explicit consideration of no-trade intervals and trade size effects as well as history dependence of the behavior of uninformed traders. | MLE based on 5 different trading events, assuming that the transition probabilities between different trading events follow information specific first order Markov chains. | Daily trade event counts for six stocks traded on the NYSE from the ISSM database for the period October 1 until December 22, 1990 (60 trading days). | Both, trade size and no-trade intervals are informative, when history dependence of uninformed behavior is allowed for. The evidence on the information content of trade size holds for all but one of the stocks considered, while uninformed trading appears to be history dependent and no-trade intervals appear to be informative in all cases. |
| Easley, O'Hara, and Paperman (1998) | Basic STM. | Same as in Easley, Kiefer, and O'Hara (1996). | Daily trade event counts for 120 stocks traded on the NYSE from the ISSM database for the period of October 1 until December 22, 1991 (60 trading days). The number of financial analysts, that follows a particular stock, is from the I/B/E/S database. | PIT is lower for stocks followed by a high number of analysts. The low analyst group of stocks tends to have lower arrival rates of both, informed and uninformed traders. The number of analysts is negatively related to private information, but positively related to the overall level of trading. |
| Easley, O'Hara, and Srinivas (1998) | STM with explicit consideration of dual trading on stock and option markets. | Indirect test of STM implications based on time series regressions and Granger causality tests. | Intradaily time series of stock prices and option volumina traded on the NYSE and options traded on the CBOE for a sample of 50 stocks from the ISSM database for the period between October 1 until November 30, 1990 (44 trading days). | Option volumes have predictive power for stock price changes. Informed traders trade on both, option and stock markets. |
| Heidle and Huang (1999) | Basic STM. | Same as in Easley, Kiefer, O'Hara, and Paperman (1996). | Daily trade event counts for 96 stocks that switched between NYSE, NASDAQ and AMEX in 1996 from the TAQ database. The sample period spans 75 trading days before and 75 trading days after the listing on the new exchange. | PIT and bid-ask spreads drop when firms move from NASDAQ to NYSE or AMEX, NASDAQ and AMEX do not change when they move from AMEX to NYSE. The spread is an increasing function of PIT. |

**Table 3.1:** Empirical research using sequential trade models (cont.)

| Authors | Model-type | Estimation method | Data set | Main results |
|---|---|---|---|---|
| Grammig, Schiereck, and Theissen (1999) | Basic STM. | Same as in Easley, Kiefer, and O'Hara, and Paperman (1996). | Daily trade event counts for the 30 DAX stocks that are contained in the IBIS and traded on the IBIS for the period January 1 until March 31, 1995 (64 trading days). | There are no systematic differences in PIT between high and low volume stocks, but average rival rates of informed traders and average bid-ask spreads are lower for high volume stocks. |
| Brown, Thomson, and Walsh (1999) | STM with explicit consideration of no-trade intervals, trade size effects and limit order submission. | MLE based on 9 different ent trading events; otherwise identical to Easley, Kiefer, and O'Hara (1997b). | Daily and intradaily trade event and order submission counts for six stocks traded on the ASX from the SIRCA database for a period of 250 trading days prior to December 1, 1996. | Informed traders choose small volume orders more often than uninformed. There are no differences between informed and uninformed traders with respect to submission of market and limit orders. The arrival rate of informed traders is equal on good and bad news days. The arrival rates of informed and uninformed traders follow the same U-shaped intraday pattern as the time series of trades. |
| Jern (2000) | STM with explicit consideration of dual trading on stock and option markets and history dependent order arrival rates. | MLE based on 8 different ent trading events; otherwise identical to Easley, Kiefer, and O'Hara (1997a). | Daily trade event counts for 6 stocks traded on the XETRA, and corresponding options tradaily on the DTB for the period December 8, 1997 until February 18, 1998 (49 trading days). | Informed traders prefer stock over options markets. Feedback effects between markets are driven by uninformed traders learning behavior, as well as hedging and arbitrage considerations. |
| Easley, O'Hara, and Saar (2001) | Extension of basic STM with explicit consideration of wise identical to limit order submission by uninformed man (1996). | MLE based on 4 different ent trading events; otherwise identical to Easley, O'Hara, and Paper- Kiefer, | Daily trade event counts for the 72 other- stocks traded on the NYSE that had creases after a stock split, a 2 to 1 stock split in 1995 from the TAQ database. All trades that occurred in the period beginning 20 days before the announcement of the split and ending 20 days after the stock split were eliminated. The sample period spans 45 trading days before and 45 trading days after the eliminated period. | Informed and uninformed trading activity increases after a stock split, while PIT decreases very slightly. The execution rates of limit orders increase after the split. The spread rises as well, but this appears to be unrelated to informed trading. |

**Table 3.1:** Empirical research using sequential trade models (cont.)

| Authors | Model-type | Estimation method | Data set | Main results |
|---|---|---|---|---|
| Grammig, Schiereck, Theissen (2001) | Extension of basic STM with explicit consideration of parallel trading in floor and electronic markets. | MLE based on 4 different trading events; identical to Kiefer, O'Hara, and Paperman (1996). | Daily trade event counts for the 30 PIT and other stocks contained in the DAX and FSE. Easley, traded on the IBIS and FSE for the period June 1 until July 31, 1997 (44 trading days). | bid-ask spreads are higher on the PIT and IBIS than on the FSE. The spread is an increasing function of PIT. |
| Easley, Hvidkjaer, and O'Hara (2001) | Extension of basic STM, allowing for different buy and sell order arrival rates of uninformed traders. | Same as in Easley, Kiefer, and O'Hara, and Paperman (1996). | Daily trade event counts for all ordinary stocks listed on the NYSE between 1983 and 1998. The sample size varies between 1311 and 1846 firms per year. Only firms, for which observations on at least 60 trading days in a given year are available, are included. Estimation is performed for each stock and each year separately. Data is from the ISSM and TAQ databases. | The arrival rates of uninformed and informed traders follow the same upward time trend as trading volume, but PIT appears to be stable across years, both individually and cross-sectionally. Volume is negatively correlated with PIT across stocks. Generally, PIT has predictive power for expected returns, and its impact is greater for small volume than for large volume stocks. |
| Lei and Wu (2001) | Extension of basic STM, allowing for history dependent arrival rates of uninformed and informed traders. | MLE based on two trading events, with arrival rates dependent that evolve according to a two state Markov chain. | Daily trade event counts for 8 stocks listed on the NYSE from the TAQ database during the period January 1, 1993 until December 31, 2000 (2000 trading days). | The arrival rates of uninformed and informed traders are time varying. Uninformed arrival rates seem to be influenced by the evolution of market fundamentals and indicate herd trading behavior, while informed traders adjust their trading activities to the level of uninformed trading. PIT is positively related to bid-ask spreads and volatility. |

Table 3.1: Empirical research using sequential trade models (cont.)

| Authors | Model-type | Estimation method | Data set | Main results |
|---|---|---|---|---|
| Weston (2001) | Extension of basic STM, allowing information regimes; for three types of traders, uninformed liquidity and noise traders, as well as informed traders. | MLE based on 6 different wise identical to Easley, O'Hara, and Paper-man (1996). | Daily trade event counts for 4906 stocks traded on the NYSE, AMEX, and NASDAQ from the TAQ database during the first quarter of 1998 (65 trading days). | Estimates of PIT, as well as of the arrival rates of informed and liquidity traders are smaller, when noise trading is allowed for. PNT is highest for medium volume stocks, and lowest for high volume stocks. Noise traders prefer NASDAQ over NYSE and AMEX. |
| Easley, Engle, O'Hara, and Wu (2002) | Extension of basic STM, allowing for time varying order arrival rates of uninformed and informed traders. | MLE, based on 2 different trade event counts. | Daily trade event counts for 16 actively traded stocks that are listed on the NYSE from the TAQ database during the period January 3, 1983 until December 24, 1998 (3891 trading days). | The arrival rates of informed and uninformed traders have an upward time trend and tend to be auto- and crosscorrelated. Volatility appears to be an increasing function of the trade arrival rates, but unrelated to PIT. The bid-ask spread tends to be negatively correlated with PIT. The increase of the total trading volume during the sample period is associated with an increase of market depth, but negatively correlated with the predicted arrival rates of informed traders. |
| Hujer, Vuletić, and Kokot (2002) | Extension of basic STM, allowing for different buy and sell order arrival rates of uninformed and informed traders. | Indirect test of STM implications based on a 3 regime and MSACD model for the time series of durations between arrival of uninformed trades, assuming a conditional Burr distribution for the durations, given the information regime. | Intradaily duration data for one stock (Boeing) traded on the NYSE from the TAQ database for the period November 1 until November 27, 1996 (19 trading days). | On a trade-by-trade basis, the relative magnitudes of the (time-varying) regime probabilities derived from the estimated MSACD model have significant predictive power for the inclination to buy the stock. |

**Table 3.1:** Empirical research using sequential trade models (cont.)

| Authors | Model-type | Estimation method | Data set | Main results |
|---|---|---|---|---|
| Fu (2002) | Basic STM. | Same as in Easley, Kiefer, and O'Hara, and Paperman (1996). | Daily trade event counts for 94 NYSE-listed firms that undertook an equity carve-out and an equally sized control sample from the TAQ database between 1991 and 2001. The sample period spans 45 trading days before the announcement date and 45 trading days after the offer date. | Informed and uninformed trade arrival rates increase after the equity carve-out. The probability of a news event and PIT decrease. In the control sample no systematic change of PIT occurs. |
| Aktas, de Bodt, Declerck, and Van Oppens (2003) | Basic STM. | Same as in Easley, Kiefer, and O'Hara, and Paperman (1996). | Daily trade event counts for 141 stocks involved in mergers and acquisitions traded on the Paris stock exchange between 1995 and 2000 from the Euronext database. The sample period spans up to 270 trading days before and 63 days after the announcement date. | Estimates of PIT increase after the announcement date. |
| Beber and Caglio (2002) | Basic STM. | Same as in Easley, Kiefer, and O'Hara, and Paperman (1996). | Daily trade event counts for 10 stocks traded on the NYSE from the TORQ database for the period from November 1990 through January 1991 (21 trading days). | Estimates of PIT are used to categorize stocks. In the high PIT group of stocks the frequency of passive limit orders is higher (21 than in the low PIT group of stocks. |

# 4

# Econometric Analysis of Sequential Trade Models

## 4.1 The EKOP Model and Finite Mixture Models

### 4.1.1 Motivation

In this Chapter we will motivate the derivation of the sample log likelihood for several generalizations of the model developed by Easley, Kiefer, O'Hara, and Paperman (1996). These models will also prove useful, when it comes to testing some of the restrictions of the EKOP model, that are likely to be violated in many samples, while the basic structure of the approach still holds. As an illustration, consider the case, in which the arrival rate of informed sellers on bad news days differs systematically from the arrival rate of informed buyers on good news days, which in the EKOP framework is restricted to be the same. This might be due to short selling restrictions or stem from credit market imperfections that lead to an inability of the informed traders to capitalize their superior information completely.[1]

In addition, the behavior of uninformed traders may well depend on the type of day, e.g. if we introduce the possibility of *strategic behavior* by uninformed traders.[2] In this case uninformed traders might want to mimic the actions of possibly better informed traders by conditioning their behavior on the observable order flow. Thus, e.g. a higher rate of sell order arrivals initially caused by informed traders during a bad news day will induce additional sells by uninformed as they try to learn from the observed order flow. As a result of such behavior, the arrival rate of uninformed traders may also depend on the type of information event, even though uninformed traders do not observe the signal themselves. It is obvious, that the implications of the original variant of

---

[1] See Diamond and Verrecchia (1987).

[2] See Admati and Pfleiderer (1988), Admati and Pfleiderer (1989), Easley, Kiefer, and O'Hara (1997a), Brown, Thomson, and Walsh (1999), and Lei and Wu (2001) for microstructure models that allow uninformed behavior to be related to the observable order flow.

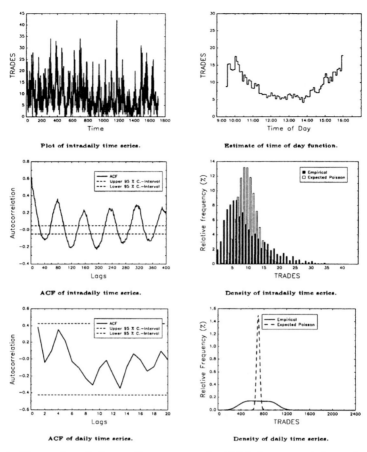

**Fig. 4.1.** Properties of the trading process for IBM shares traded on the NYSE during August 1996. The time of day function has been estimated by the average number of trades over successive 5 minute intervals. The expected density is the Poisson evaluated using the sample means of 8.98 trades per 5 minute interval respectively 700.55 trades per day.

the EKOP model will not hold in such a case, as the possibility of uninformed herd trading is excluded a priori.

Our primary goal however is to use the information contained in available high frequency data sets more efficiently than has been done in previous research. As laid out in Chap. 3, most empirical studies estimate structural parameters of the EKOP model (or variants thereof) based on samples of daily trading event counts. The daily sampling scheme implies a loss of efficiency in estimation, since the number of observations drops significantly, compared to estimation based on a sample of observations for some intradaily interval.

For example, trading on the NYSE starts every day at 9.30 a.m. and ends at 16.00 p.m., so the daily sampling scheme aggregates all trades over an interval of 6.5 hours. When data is sampled over an interval of 5 minutes there are 78 observations for each day instead of only one. Consider using a data set that spans a period of 60 days, which is the typical horizon in many studies, then the total number of observations drops from 4680 using the 5-minute sampling scheme to 60 at the daily level.

Apart from the efficiency issue, there are many interesting features of the trading process, that cannot be examined at all using samples of daily observations. For example, it is a well documented stylized fact, that transaction arrival rates on many financial markets have a characteristic intradaily seasonal pattern and time series of durations between successive trades tend to be autocorrelated.[3] These features of the trading process carry over to the event counts used to estimate the EKOP model as exemplified by the data gathered in Fig. 4.1 for the total number of trades of the IBM stock. Time series of other trading events (e.g. buys and sells) sampled over intradaily time intervals for several stocks listed on the NYSE display similar patterns.[4]

The deterministic seasonal pattern of course disappears if the data are sampled at the daily frequency, as has been done by EKOP. Furthermore, when sampled at the daily frequency, the serial correlation of the trades is reduced considerably. In addition, the empirical density of the daily trades is far from being even similar to a Poisson. In fact, due to the increased range of the observed values caused by the aggregation of trades on the daily level[5], one rarely finds any two coincident values in a sample.

These patterns may result from strategic behavior of traders or might be caused by other factors that are not addressed explicitly by the EKOP model. A primary question of interest is then whether uninformed and informed traders behavior differs with respect to the time of day, at which they prefer to trade and also whether their reaction function to the observable trade flow exhibits significant differences. An alternative explanation for the seasonality is, that these patterns may be related to institutional features of the trading process, e.g. to the periodic opening and closure of other asset markets. Moreover they might be caused by habits, e.g. the general decrease of trading intensity during lunch time. In this case we would not expect to

---

[3] See e.g. Engle and Russell (1998), Hasbrouck (1999), and Grammig, Hujer, and Kokot (2002).

[4] See Sect. 5.3 for details.

[5] The minimum observed number for the trades is 351, while the maximum is equal to 1051. Since the distribution of the data can hardly be visualized by a histogram of event counts under these circumstances, we computed a kernel density estimate and plotted it against a Poisson probability function in the lower right panel of Fig. 4.1. We used the sample mean of 700.55 trades as an estimate for the mean parameter of the Poisson distribution. The empirical density was estimated employing the Gaussian kernel with bandwidth parameter chosen according to Silverman's rule of thumb, see Silverman (1986) for details.

observe significant differences between informed and uninformed traders. In the following Sections of this Chapter we will introduce econometric methods, that may be used to address these issues explicitly.

### 4.1.2 An Alternative Version of the EKOP Model

In a first step towards a more general model, that is able to describe these well known empirical phenomena, we will ignore these features of the data. We will present enhanced versions of the EKOP model in the following Sections, that may be able to account for intradaily variation of arrival rates as well as for serial correlation. In the presence of these phenomena, estimates of structural parameters that result from the procedure proposed by EKOP for intradaily sampling schemes may be severely biased, since such departures from the model postulates are not accounted for and in most empirical applications of the EKOP model and its variants the validity of the assumptions is not even tested for appropriately.

In order to gain some insights into the problems associated with the estimation of models that incorporate these features we will now treat the EKOP model explicitly as a mixture model for a multivariate poisson process, and derive the corresponding likelihood function. This will prove useful when we introduce further extensions of the basic approach. We will call this version of the model the *ignorant agnostic model* in order to distinguish it from the original EKOP model. Indeed, if we restrict our attention to an agnostic version of the EKOP model and allow all arrival rates in the model to be distinct and at the same time ignore systematic intradaily trading patterns and time dependence, the structure of the trading process may be summarized by the tree diagram in Fig. 4.2. The ignorant agnostic model therefore nests the original EKOP model as a special case.

**Fig. 4.2.** Structure of the *ignorant agnostic* trading model.

This model can be described as a mixture of Poisson distributions with mixture probabilities $\pi_j$ that correspond to the *unconditional probability* of the unobserved, discrete regime variable $r_t$ to be equal to state $j$, i.e.

$$p\left(r_t = j; \theta\right) = \pi_j, \tag{4.1}$$

where $t = 1, 2, \ldots, T$ denotes the trading interval[6], $j = 1, 2, \ldots, N$ is the number of states and

$$\theta = \left(\lambda_{11}, \ldots, \lambda_{N1}, \lambda_{12}, \ldots, \lambda_{NK}, \pi_1, \ldots, \pi_N\right)'$$

is the vector of parameters if we distinguish between $K$ different trading events.[7] The regime variable $r_t$ corresponds to the information event at the beginning of each observation interval. The EKOP model has $N = 3$ different states and distinguishes between $K = 2$ different trading events (buys resp. sells).

The *conditional marginal density* of the counts of the $k$-th trading event $y_{tk}$ for an arbitrary trading interval $t$ of length $\Delta t$, taking $r_t$ as given is

$$f_{y_{tk}}\left(y_{tk} \mid r_t = j; \theta\right) = \frac{\exp\left(-\lambda_{jk} \cdot \Delta t\right) \cdot \left(\lambda_{jk} \cdot \Delta t\right)^{y_{tk}}}{y_{tk}!}. \tag{4.2}$$

We still assume, that arrivals of informed and uninformed traders are, conditionally on the regime variable $r_t$, independently distributed, so the *conditional joint distribution* of observing a realization of the vector of observations $y_t = \left(y_{t1}, \ldots, y_{tK}\right)'$ is just the product of the conditional marginal densities

$$f_j\left(y_t \mid r_t = j; \theta\right) = \prod_{k=1}^{K} f_{y_{tk}}\left(y_{tk} \mid r_t = j; \theta\right). \tag{4.3}$$

Since we cannot observe the value of the state variable, the likelihood function has to be derived from the *unconditional joint density* of the observations, which is equal to the sum of the joint densities $g_j\left(\cdot\right)$ of the random variables $y_t$ and $r_t$ over all possible values of the regime variable $r_t$

$$f_t\left(y_t; \theta\right) = \sum_{j=1}^{N} g_j\left(y_t \cap r_t = j; \theta\right) \tag{4.4}$$

$$= \sum_{j=1}^{N} \pi_j \cdot f_j\left(y_t \mid r_t = j; \theta\right),$$

---

[6] In order to keep the exposition as general as possible, we will not distinguish between the time of day index $\tau$ and the number of day index $d$ in the following. Thus, the trading interval can either be a complete trading day, or some intradaily interval of fixed length.

[7] Of course only $N - 1$ mixture probabilities can be treated as free parameters, since $\pi_N = 1 - \sum_{i=1}^{N-1} \pi_i$. In order to keep the exposition as simple as possible, we will stick to the notation introduced above.

$$= \sum_{j=1}^{N} \pi_j \cdot \prod_{k=1}^{K} f_{y_{tk}} \left( y_{tk} \mid r_t = j; \theta \right).$$

Thus, the log likelihood function for the observed data, also known as *incomplete* log likelihood $\mathcal{L}_I(\theta)$, can be calculated from (4.4) as

$$\ln \mathcal{L}_I (\theta) = \sum_{t=1}^{T} \ln f_t (y_t; \theta). \tag{4.5}$$

Estimates of $\theta$ can be obtained by maximizing (4.5) subject to the constraint

$$\sum_{j=1}^{N} \pi_j = 1.$$

If we introduce the *conditional density* of the regime variable $r_t$ given the data $y_t$, which is equal to

$$p_t (r_t = j \mid y_t; \theta) = \frac{\pi_j \cdot f_j (y_t \mid r_t = j; \theta)}{f_t (y_t; \theta)}, \tag{4.6}$$

the maximum likelihood estimates can be characterized as the solution of the following set of nonlinear equations[8]

$$\widehat{\lambda}_{jk} = (\Delta t)^{-1} \cdot \frac{\sum\limits_{t=1}^{T} y_{tk} \cdot p_t (r_t = j \mid y_t; \theta)}{\sum\limits_{t=1}^{T} p_t (r_t = j \mid y_t; \theta)}, \tag{4.7}$$

and

$$\widehat{\pi}_j = T^{-1} \cdot \sum_{t=1}^{T} p_t (r_t = j \mid y_t; \theta). \tag{4.8}$$

The set of equations (4.7) and (4.8) may be used to estimate the parameters of the model in a very convenient way, employing a variant of the EM-algorithm.[9] One proceeds in the following manner:[10] Starting from an arbitrary initial value for the parameter vector $\theta^{(0)}$, one calculates the conditional probabilities in (4.6). These can be used together with $\theta^{(0)}$ to evaluate the right side of (4.7) and (4.8). The left sides of (4.7) and (4.8) then produce a new value $\theta^{(1)}$. It can be shown, that during each iteration of this algorithm the value of the likelihood function is increased. Thus, continuing in this fashion until the maximum absolute change of the parameter vector $\max \mid \theta^{(p+1)} - \theta^{(p)} \mid$ is smaller than some prespecified convergence criterion $\epsilon$, one obtains the

---

[8] See Appendix A.2 for a derivation.
[9] See Appendix A.3 for an exposition of the EM-algorithm.
[10] See Hamilton (1994), p. 688.

maximum likelihood estimates of $\theta$. The use of the EM-algorithm for the estimation of mixture models is highly recommended in the literature, because of its outstanding stability compared to other iterative search methods.[11]

In effect we have derived a reduced form sequential trade model. The parameter $\lambda_{jk}$ represents the joint arrival rate of informed and uninformed traders for the $k$-th trading event under regime $j$. Individual arrival rates of informed and uninformed traders cannot be identified unambiguously in our framework, but the nature of the information regime may be determined by comparing the regime specific joint arrival rates. If e.g. the sell arrival rate under regime $j$ is significantly larger than the corresponding arrival rate of buys, we may conclude that state $j$ is the bad news regime. Thus, the nature of the regime may be assessed post estimation by comparing estimates of $\lambda_{jk}$.[12]

The estimation of reduced form models has the additional advantage, that the number of mixture components $N$ can be determined on the basis of statistical criteria[13], while in structural models the number of information regimes has to be imposed a priori and cannot be tested in any obvious manner. Consider for example the case of the original EKOP model, that has three information regimes. In the good news regime the buy arrival rate $\varepsilon + \mu$ is higher than the sell arrival rate $\varepsilon$. In the bad news regime the sell arrival rate $\varepsilon + \mu$ will be higher than the buy arrival rate $\varepsilon$, while under no news both arrival rates will be equal to $\varepsilon$. If we want to test this model e.g. against a four regime alternative, it is unclear how the arrival rates under the fourth regime should be specified. The results of statistical tests may therefore lead to conflicting conclusions depending on what restrictions will be imposed on the arrival rates in the forth regime.

When using a reduced form model we will simply estimate two more arrival rates $\lambda_{41}$, and $\lambda_{42}$ and check whether such a specification will improve the fit of the model to the data without imposing any restriction on whether the arrival rate of buys is higher or lower than that for the sells under the fourth regime. Furthermore, the observed data might as well be consistent with a two regime specification rather than a three regime model. Apart from the missing economic content of a sequential trade model allowing for e.g. a good and a no news regime only, statistical inference may again be affected in nontrivial ways by dropping an arbitrary regime. Thus, there is a problem of indeterminacy in the structural approach, that prevents us from assessing the appropriate number of regimes in a manner consistent with the underlying theory. This problem can be avoided by the use of reduced form models.

### 4.1.3 A Multivariate Finite Mixture Poisson Regression Model

If we want to include explanatory variables, e.g. in order to control for intradaily seasonal patterns, we have to generalize the framework of the last

---

[11] See Böhning (1995) for a discussion.
[12] This feature of our model is similar to the framework of Lei and Wu (2001).
[13] We will discuss such criteria in Sect. 4.2.2 below.

Section slightly. The following model is a multivariate version of the latent class Poisson regression model of Wedel, Desarbo, Bult, and Ramaswamy (1993). We still assume that the unconditional probability of the regime variable $r_t$ is given by (4.1), while the conditional density of the counts of the $k$-th trading event $y_{tk}$ is now governed by a Poisson regression model[14] with conditional mean given by

$$E\left(y_{tk} \mid r_t = j, \mathcal{F}_t; \theta\right) = \lambda_{tjk} \cdot \Delta t = \exp\left(x'_t \beta_{jk}\right) \cdot \Delta t, \tag{4.9}$$

where $\mathcal{F}_t$ denotes the information set that is available at time $t$ and may include lagged dependent as well as contemporaneous and lagged exogenous variables, $\lambda_{tjk}$ denotes as before the mean arrival rate per unit of time, $x_t = (x_{t1}, \ldots, x_{tM})'$ is a $(M \times 1)$ vector of explanatory variables at time $t$, and $\beta_{jk} = (\beta_{jk1}, \ldots, \beta_{jkM})'$ is the corresponding parameter vector of the conditional mean for the variable $y_{tk}$, given regime $r_t = j$. Note that (4.1) implies that $r_t$ is independent of $x_t$. Thus the vector of parameters for $k$ trading events, given $N$ different states is

$$\theta = (\beta_{11}, \ldots, \beta_{N1}, \beta_{12}, \ldots, \beta_{NK}, \pi_1, \ldots, \pi_N)'.$$

The structure of this generalized sequential trade model is reproduced in Fig. 4.3.

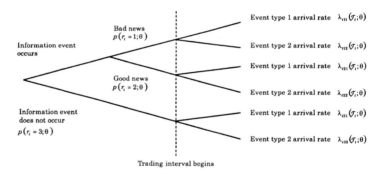

Fig. 4.3. Structure of the generalized sequential trade model.

We will estimate parameters employing the EM-algorithm. Therefore, we have to derive the complete data likelihood $\mathcal{L}_C(\theta)$, i.e. the likelihood function that would be appropriate, if the regime variable $r_t$ was directly observable. This can be derived from the joint densities $g_j(\cdot)$ of the random variables $y_t$ and $r_t$. In order to present the likelihood in a compact notation, we introduce the random variable $\xi_t^{(j)} = I(r_t = j)$ where $I(r_t = j)$ denotes the indicator

---

[14] See Appendix A.4 for an exposition of the Poisson regression model.

function, which is equal to one, if $r_t = j$ and zero otherwise. With this in hand $\mathcal{L}_C(\theta)$ is given by

$$\mathcal{L}_C(\theta) = \prod_{t=1}^{T} \prod_{j=1}^{N} \prod_{k=1}^{K} f_{y_{tk}}\left(y_{tk} \mid r_t = j, x_t; \theta\right)^{\xi_t^{(j)}} \cdot \pi_j^{\xi_t^{(j)}}, \tag{4.10}$$

with associated log likelihood

$$\ln \mathcal{L}_C(\theta) = \sum_{t=1}^{T} \sum_{j=1}^{N} \sum_{k=1}^{K} \xi_t^{(j)} \cdot \ln f_{y_{tk}}\left(y_{tk} \mid r_t = j, x_t; \theta\right) \tag{4.11}$$

$$+ \sum_{t=1}^{T} \sum_{j=1}^{N} \xi_t^{(j)} \cdot \ln \pi_j.$$

A typical iteration of the EM-algorithm is then defined by two steps, the *expectation* step (E-step), in which the conditional expectation of the complete data log likelihood $\ln \mathcal{L}_{EC}(\theta) \equiv E[\ln \mathcal{L}_C(\theta) \mid \mathcal{Y}_T; \theta^{(p)}]$ given the observable data $\mathcal{Y}_T = (y_1, \ldots, y_T, x_1, \ldots, x_T)$ and the current guess of the parameter vector $\theta^{(p)}$ is formed and the *maximization* step (M-step), in which the expected log likelihood is maximized with respect to $\theta$ in order to obtain the next provisional guess $\theta^{(p+1)}$. In our model the E-step is equivalent to replacing the unobserved variable $\xi_t^{(j)}$ by its expectation $\hat{\xi}_t^{(j)}$, given the data and $\theta^{(p)}$. This may be evaluated by using[15]

$$\hat{\xi}_t^{(j)} = \frac{\pi_j \cdot f_j\left(y_t \mid r_t = j, x_t; \theta\right)}{f_t\left(y_t \mid x_t; \theta\right)}. \tag{4.12}$$

After replacing $\xi_t^{(j)}$ by $\hat{\xi}_t^{(j)}$ in (4.11), the M-step can be conducted. From (4.11) we see that the maximization may be separated with respect to the regime probabilities $\pi_j$ and each of the vectors of regression coefficients $\beta_{jk}$. Maximization with respect to $\pi_j$ can be performed by maximizing the augmented function[16]

$$J(\pi) = \sum_{t=1}^{T} \sum_{j=1}^{N} \hat{\xi}_t^{(j)} \cdot \ln \pi_j + \gamma \left(1 - \sum_{j=1}^{N} \pi_j\right), \tag{4.13}$$

where $\gamma$ denotes the Lagrange multiplier corresponding to the restriction $\sum_{j=1}^{N} \pi_j = 1$.

Setting the derivative of (4.13) with respect to $\pi_j$ equal to zero, solving for $\gamma$, and rearranging yields[17]

---

[15] Note the analogy to (4.6).

[16] This is equivalent to forming the Lagrangean with respect to the restriction $\sum_{j=1}^{N} \pi_j = 1$ from the complete data log likelihood (4.11) as shown in Appendix A.2 for the case of the ignorant agnostic model.

[17] This is completely analogous to the derivation of (4.8).

$$\hat{\pi}_j = \frac{1}{T} \cdot \sum_{t=1}^{T} \hat{\xi}_t^{(j)}. \qquad (4.14)$$

Maximization with respect to the regression coefficients $\beta_{jk}$ can be carried out by maximizing independently each of the $(K \times N)$ expressions

$$\ell_{jk} = \sum_{t=1}^{T} \hat{\xi}_t^{(j)} \cdot \ln f_{y_{tk}} \left( y_{tk} \mid r_t = j, x_t; \theta \right). \qquad (4.15)$$

The derivatives are given by

$$\frac{\partial \ell_{jk}}{\partial \beta_{jk}} = \sum_{t=1}^{T} \hat{\xi}_t^{(j)} \cdot (y_{tk} - \lambda_{tjk} \cdot \Delta t) \cdot x_t, \qquad (4.16)$$

which is equal to the derivative condition of the ordinary Poisson regression model weighted by the regime probabilities $\hat{\xi}_t^{(j)}$. Thus, the estimates of the regression parameters can be obtained for each of the $K$ equations and each of the $N$ regimes separately by a weighted Poisson regression.

Starting with an arbitrary first guess $\theta^{(0)}$ the EM-algorithm usually converges quickly to a local maximum of the complete log-likelihood function which is at the same time a local maximum of the incomplete log-likelihood. As shown in Appendix A.3, standard errors of the parameter estimates can be derived from the derivatives and second derivatives of the expected log likelihood $\ln \mathcal{L}_{EC}(\theta)$. In the case of the model described above, it turns out, that both, the derivatives and the Hessian matrix of the regression parameters $\beta_{jk}$ are simple functions of the corresponding expressions of the ordinary Poisson regression model, that has been used to conduct the maximization step. Since $\frac{\partial \ln \mathcal{L}_{EC}(\theta)}{\partial \beta_{jk}} = \frac{\partial \ell_{jk}}{\partial \beta_{jk}}$, an expression for the gradient has already been given by (4.16) and the Hessian is

$$\frac{\partial^2 \ln \mathcal{L}_{EC}(\theta)}{\partial \beta_{jk} \partial \beta'_{jk}} = - \sum_{t=1}^{T} \hat{\xi}_t^{(j)} \cdot \lambda_{tjk} \cdot \Delta t \cdot x_t x'_t. \qquad (4.17)$$

With respect to the regime probabilities $\pi_j$, it can be shown, that the gradient is given by

$$\frac{\partial \ln \mathcal{L}_{EC}(\theta)}{\partial \pi_j} = \sum_{t=1}^{T} \left( \frac{1}{\pi_j} \cdot \hat{\xi}_t^{(j)} - 1 \right), \qquad (4.18)$$

while the Hessian is equal to

$$\frac{\partial^2 \ln \mathcal{L}_{EC}(\theta)}{\partial \pi_j^2} = - \sum_{t=1}^{T} \left( \frac{1}{\pi_j^2} \cdot \hat{\xi}_t^{(j)} \right). \qquad (4.19)$$

The cross derivatives $\frac{\partial^2 \ln \mathcal{L}_{EC}(\theta)}{\partial \beta_{jk} \partial \pi_j}$ as well as $\frac{\partial^2 \ln \mathcal{L}_{EC}(\theta)}{\partial \beta_{jk} \partial \beta'_{il}}$, $\frac{\partial^2 \ln \mathcal{L}_{EC}(\theta)}{\partial \beta_{jk} \partial \beta'_{jl}}$ and $\frac{\partial^2 \ln \mathcal{L}_{EC}(\theta)}{\partial \pi_j \partial \pi_i}$ for $i \neq j$ and $k \neq l$ are all equal to zero. Thus, in order to derive the standard errors for the parameter vector $\theta$, the usual estimates of the information matrix $\mathcal{I}(\theta)$ can be computed employing expressions (4.16) through (4.19) and estimates of $Cov(\theta)$ may be obtained through $Cov(\hat{\theta}) = T^{-1} \cdot \mathcal{I}(\hat{\theta})^{-1}$.

### 4.1.4 A Mixture Regression Model Based on the Negative Binomial Distribution

In many applications of count data regression analysis it is recognized that the simple Poisson regression model is too restrictive to explain certain features of the data. Often the observed sample data exhibits *overdispersion*, i.e. the sample variance is greater than the mean. This is at odds with the assumption, that the data generating process is a Poisson model, since the Poisson distribution has mean equal to the variance. In such cases alternative distributions that allow for overdispersion like the negative binomial may be employed.[18] In the context of this study, overdispersion may arise as a consequence of unobserved heterogeneity among the population of traders. Note, that we are not able to observe individual characteristics of the market participants, but only the outcomes of the trading process as a result of the interaction of a possibly heterogeneous population of traders.

In the original EKOP framework, it is assumed that there is only heterogeneity between the two groups of informed and uninformed traders, i.e. *between-group variation*, leading to group specific trade arrival rates, but individual members of each group all have the same arrival rate, thus there is no *in-group variation* by assumption. However, it may be more realistic to assume that traders in each group will have different individual specific characteristics, e.g. with respect to their endowments, or their attitude towards risk and that these characteristics influence their trading behavior.[19] A mixture of Poisson distributions can be used to model between-group variation, since it has a conditional variance that is greater than its conditional mean[20] but it will fail to account for in-group variation.

Consider instead the situation, when the population of market participants consists of a set of traders, each associated with an individual trading intensity $\lambda_s$ that can be described by a Poisson process and the trading intensities of different individuals are equal up to a multiplicative shift, i.e. the trading intensity of trader $s$ is given by $\lambda_s = \psi_s \cdot \lambda$. If we assume, that $\psi_s$ is a Gamma distributed random variable with mean equal to one and variance equal to $\alpha$, it can be shown that the marginal distribution of the trade event counts will be a negative binomial distribution with arrival rate given by the population mean of the individual arrival rates $\lambda = E(\lambda_s)$ and dispersion parameter equal

---

[18] See Appendix A.5 for an exposition of the ordinary negative binomial regression model.

[19] See Saar (2000) and Saar (2001a) for an assessment of these issues in an enhanced sequential trade framework.

[20] See Appendix A.6.

to the variance of the individual specific shift parameters $\alpha = Var(\psi_s)$.[21] In the context of our study it seems reasonable that the composition of the actively participating population of traders changes systematically with the information regime and the type of the trading event, hence we allow for regime and event specific heterogeneity parameters $\alpha_{jk}$.

Thus, accounting for in-group variation yields an alternative version of the mixture regression model of the last Section that can be implemented by replacing the conditional density in (4.2) with

$$
f_{y_{tk}}\left(y_{tk} \mid r_t = j; \theta\right) = \frac{\Gamma\left(y_{tk} + \alpha_{jk}^{-1}\right)}{\Gamma\left(y_{tk} + 1\right) \cdot \Gamma\left(\alpha_{jk}^{-1}\right)}
$$

$$
\cdot \left(\frac{\alpha_{jk}^{-1}}{\alpha_{jk}^{-1} + \Delta t \cdot \lambda_{tjk}}\right)^{\alpha_{jk}^{-1}}
$$

$$
\cdot \left(\frac{\Delta t \cdot \lambda_{tjk}}{\alpha_{jk}^{-1} + \Delta t \cdot \lambda_{tjk}}\right)^{y_{tk}}, \tag{4.20}
$$

where $\Gamma(s) = \int_0^\infty \exp(-u) \cdot u^{s-1} du$ denotes the Gamma function. (4.20) is the density function of the negative binomial distribution with parameters $\lambda_{tjk}$ and $\alpha_{jk}$. The parameterization above is commonly referred to as the Negbin 2 model. As before, we assume that $\lambda_{tjk}$ is determined by the log-linear specification given in (4.9), so the vector of parameters changes to

$$
\theta = (\beta_{11}, \ldots, \beta_{NK}, \alpha_{11}, \ldots, \alpha_{NK}, \pi_1, \ldots, \pi_N)'.
$$

The Negbin 2 specification is appropriate to capture possible regime specific heterogeneity, since the regime specific conditional variance function $v_{tjk}$ is given by $v_{tjk} = \lambda_{tjk} + \alpha_{jk} \cdot \lambda_{tjk}^2$ instead of $v_{tjk} = \lambda_{tjk}$ as in the equidispersed Poisson model.

The EM-algorithm proposed in the last Section can be employed to estimate this model as well. The evaluation of the expectation of the unobserved regime probabilities $\xi_t^{(j)}$ proceeds as laid out in (4.12) with the Poisson distribution replaced by the Negbin 2. In the M-step, estimates of the regime probabilities can be obtained as in (4.14), though the derivative condition given in (4.16) changes to the derivative conditions of the ordinary Negbin 2-model, again weighted by the regime probabilities $\hat{\xi}_t^{(j)}$. Thus, the derivative of $\ln \mathcal{L}_{EC}(\theta)$ with respect to the regression parameter vector $\beta_{jk}$ is given by

$$
\frac{\partial \ln \mathcal{L}_{EC}(\theta)}{\partial \beta_{jk}} = \sum_{t=1}^{T} \hat{\xi}_t^{(j)} \cdot \left(\frac{y_{tk} - \lambda_{tjk} \cdot \Delta t}{1 + \alpha_{jk} \cdot \lambda_{tjk} \cdot \Delta t}\right) \cdot x_t, \tag{4.21}
$$

---

[21] See Appendix A.7 for a derivation of this result.

and the derivative with respect to $\alpha$ is

$$
\frac{\partial \ln \mathcal{L}_{EC}(\theta)}{\partial \alpha_{jk}} = \sum_{t=1}^{T} \hat{\xi}_t^{(j)} \cdot \left[ \frac{1}{\alpha_{jk}^2} \ln (1 + \alpha_{jk} \cdot \lambda_{tjk} \cdot \Delta t) \right. \tag{4.22}
$$

$$
+ \sum_{j=0}^{y_{tk}-1} \frac{j}{j \cdot \alpha_{jk} + 1}
$$

$$
\left. - \frac{\lambda_{tjk} \cdot \Delta t \cdot \left( y_{tk} + \frac{1}{\alpha_{jk}} \right)}{1 + \alpha_{jk} \cdot \lambda_{tjk} \cdot \Delta t} \right].
$$

The corresponding elements of the Hessian matrix are given by

$$
\frac{\partial^2 \ln \mathcal{L}_{EC}(\theta)}{\partial \beta_{jk} \, \partial \beta_{jk}'} = - \sum_{t=1}^{T} \hat{\xi}_t^{(j)} \cdot \frac{(1 + \alpha_{jk} \cdot y_{tk}) \cdot \lambda_{tjk} \cdot \Delta t}{(1 + \alpha_{jk} \cdot \lambda_{tjk} \cdot \Delta t)^2} \cdot x_t \cdot x_t', \tag{4.23}
$$

$$
\frac{\partial^2 \ln \mathcal{L}_{EC}(\theta)}{\partial \alpha_{jk}^2} = - \sum_{t=1}^{T} \hat{\xi}_t^{(j)} \cdot \left[ \sum_{j=0}^{y_{tk}-1} \left( \frac{j}{j \cdot \alpha_{jk} + 1} \right)^2 \right. \tag{4.24}
$$

$$
+ \frac{2}{\alpha_{jk}^3} \cdot \ln (1 + \alpha_{jk} \cdot \lambda_{tjk} \cdot \Delta t)
$$

$$
- \frac{2 \cdot \lambda_{tjk} \cdot \Delta t}{\alpha_{jk}^2 \cdot (1 + \alpha_{jk} \cdot \lambda_{tjk} \cdot \Delta t)}
$$

$$
\left. - \frac{\left( y_{tk} + \frac{1}{\alpha_{jk}} \right) \cdot (\lambda_{tjk} \cdot \Delta t)^2}{(1 + \alpha_{jk} \cdot \lambda_{tjk} \cdot \Delta t)^2} \right],
$$

$$
\frac{\partial^2 \ln \mathcal{L}_{EC}(\theta)}{\partial \beta_{jk} \, \partial \alpha_{jk}} = - \sum_{t=1}^{T} \hat{\xi}_{tj} \cdot \frac{(y_{tk} - \lambda_{tjk} \cdot \Delta t) \cdot \lambda_{tjk} \cdot \Delta t}{(1 + \alpha_{jk} \cdot \lambda_{tjk} \cdot \Delta t)^2} \cdot x_t. \tag{4.25}
$$

The standard errors of the parameters $\beta_{jk}$ and $\alpha_{jk}$ may be computed from expressions (4.21) through (4.25) in the usual manner. An univariate version of the Negbin 2 mixture model has been introduced by Trivedi and Deb (1997).

### 4.1.5 Accounting for Intraday Seasonality

It is well known that the number of trading events varies in a deterministic manner during the trading day that roughly resembles an inverted U-shaped pattern, i.e. intensity is very high after the open and before the close, while it tends to be low during the middle of the day. In order to account for the intradaily seasonal variation, we introduce a time of day function, that will be estimated by analogy to the semi-nonparametric (SNP) estimator introduced by Gallant (1981) and Eubank and Speckman (1990). The basic approach is to approximate the unknown time of day function using a fourier series expansion accommodated by polynomials in the regressor variables. Estimation is carried out by specifying a regression function of the type

$$\ln \lambda_t = \omega + \sum_{p=1}^{P} \left( \varphi_p \cdot h(\tau_t)^p \right) + \sum_{s=1}^{S} \left[ \zeta_s \cdot \cos \left( s \cdot h(\tau_t) \right) + \eta_s \cdot \sin \left( s \cdot h(\tau_t) \right) \right],$$

(4.26)

where $\tau_t$ denotes the time of day at which the $t$-th event count $y_t$ was observed, the normalizing function $h(\tau)$ is given by

$$h(\tau) = 2\pi \cdot \frac{\tau - \tau_{\min}}{\tau_{\max} - \tau_{\min}},$$

(4.27)

and $\tau_{\min}$ ($\tau_{\max}$) is the time of day at which trading begins (ends). This type of estimator is especially well suited for our purposes, because it can reproduce non-linear shapes of the time of day function. Also the SNP approach takes into account that the regressor variable has bounded support, which is true in our applications, since the trading day at the NYSE is limited to 6.5 hours per day. Asymptotic normality and consistency of SNP-estimators for several types of data generating processes with i.i.d. and heteroscedastic errors have been established in Eastwood (1991), and Andrews (1991). The same technique has been applied in several studies to estimate seasonal components in GARCH and ACD-models, see e.g. Andersen and Bollerslev (1997), or Hujer, Grammig, and Kokot (2000).

### 4.1.6 Autoregressive Specification of the Conditional Mean Function

The probably most straightforward way to account for autocorrelation in a time series of event counts is to include lagged values of the dependent variable $y_t$ in the conditional mean function. The usefulness of this specification is somewhat questionable however, especially when a log-linear conditional mean function of the form $\lambda_t = \exp(x_t'\beta)$ is specified, since the conditions for stationarity of the resulting non-linear difference equation are hard to derive. On the other hand, if a linear conditional mean function $\lambda_t = x_t'\beta$ is specified, then the non-negativity condition $\lambda_t \geq 0$ may be violated if some variables or estimates of the regression coefficients assume negative values. Also, in this specification only positive autocorrelation may be modelled, as the coefficient of $y_{t-1}$ is required to be positive as well.[22]

A more promising approach is to include lags of the logarithms of the dependent variable into the conditional mean function, since this will result in a linear autoregressive model for $\ln y_t$ for which stationarity conditions are easily derived from standard time series theory. Furthermore these conditions

---

[22] See Holden (1987) for an application of this approach.

carry over to $y_t$, i.e. if the data generating process of $\ln y_t$ is known to be stationary, then $y_t$ will be a stationary process as well. The fundamental problem when using the logarithmic variant in the count data context, is that $y_t$ may assume zero values, so that the transformation $\ln y_t$ is not defined. There are in principle two ways to circumvent this problem:[23]

The first method is to rescale only the zero values of the lags of $y_t$ to a constant $c_1$, i.e. use transformed lagged values of

$$y_t^* = \max(c_1, y_t) \tag{4.28}$$

as regressor variables, with $0 < c_1 < 1$. The second possibility is to transform all lagged values of $y_t$ by the same amount $c_2$, i.e. to use transformed lags of

$$y_t^{**} = y_t + c_2, \tag{4.29}$$

with $c_2 > 0$, as additional regressors.

The constant $c$ in both cases is arbitrarily fixed by the researcher. For example, Li (1994) estimates autoregressive count data models based on specification (4.28) with $c_1$ set to 0.5 and models based on (4.29) with $c_2$ set to 0.1. When specification (4.28) is used, the value of the parameter $c_1$ may be treated as an additional parameter to be estimated. Consider the following specification for a first order autoregressive model

$$\lambda_t^{***} = \exp\left[x_t'\beta + \phi \cdot \ln y_{t-1}^{***} + (\phi \cdot \ln c_3) \cdot d_{t-1}\right], \tag{4.30}$$

where the transformed dependent variable $y_t^{***}$ is given by

$$y_t^{***} = \begin{cases} y_t & \text{if } d_t = 0 \\ 1 & \text{if } d_t = 1 \end{cases},$$

and the dummy variable $d_t$ is defined by

$$d_t = \begin{cases} 0 & \text{if } y_t > 0 \\ 1 & \text{if } y_t = 0 \end{cases}.$$

The implied estimate of $c_3$ may be calculated by dividing the estimated coefficient of $d_{t-1}$ by the estimated coefficient of $y_{t-1}^{***}$ and taking the exponential of the result, i.e.

$$\hat{c}_3 = \exp\left(\frac{\hat{\delta}}{\hat{\phi}}\right),$$

where $\delta \equiv \phi \cdot \ln c$. Note that the conditional mean function may be rewritten as

$$\lambda_t^{***} = \exp(x_t'\beta) \cdot \begin{cases} (y_{t-1})^\phi & \text{if } y_{t-1} > 0 \\ (c_3)^\phi & \text{if } y_{t-1} = 0 \end{cases}.$$

---

[23] See Zeger and Qaqish (1988) and Cameron and Trivedi (1998), pp. 238-240.

Thus, zero values of $y_t$ are implicitly replaced by $c_3$, and the corresponding autoregressive model is well defined if $|\phi| < 1$ and may be estimated using standard Poisson regression software.[24] Extensions of this approach to higher order models are not straightforward however, since introducing separate dummy variables for each lag may produce different estimates of $\delta$ at each lag length. Also the implied estimates of $c$ may lie outside of the unit interval, thus assuming implausible values. For these reasons, it is preferable to use either specification (4.28) or (4.29) with $c$ fixed.

### 4.1.7 A Markov Switching Approach

Another possible remedy for serial dependence of the transaction time series can be found by assuming that the regime variable $r_t$ is generated by a homogenous Markov chain, i.e. that the regime variable $r_t$ follows a first order autoregressive process, that is governed by the associated transition probabilities $p_{ji}$ that relate the state of the process in the current period $r_t$ to the state of the process in past periods, i.e. $p_{ji} \equiv p(r_t = j \mid r_{t-1} = i)$ is the probability, that state $i$ in period $(t-1)$ will be followed by state $j$ in period $t$.[25]

Indeed, the original EKOP model is closely related to the class of Markov switching models in the spirit of Lindgren (1978) and Hamilton (1989). Static mixtures like the EKOP model may generally be regarded as special cases of a Markov switching model, based on independence restrictions with respect to the transition probabilities, i.e. the probabilities of moving into state $j$ in $t$ from any state $i$ in $(t-1)$ are all equal, $p_{j1} = p_{j2} = \cdots = p_{jN} \equiv \pi_j$. A natural generalization of the EKOP model is then to allow the regime probabilities to be varying in time according to a Markov chain.

This model may be particularly useful for the analysis of intradaily trading activity, as it accounts for the possibility, that informed traders may split up their transaction demand and place market orders on one side of the market over several successive trading intervals, rather than capitalizing their superior information in a single large transaction. Such behavior may cause patterns of serial correlation, and is typically not accounted for in sequential trade models. It may result from strategic considerations, e.g. by the desire of informed traders to remain unidentified.[26]

---

[24] Note, that if we would admit $c_3$ to be equal to zero, this would be an absorbing state, and once $y_{t-1}$ achieves zero, all future values of the dependent variable would be zero as well. This is an inappropriate assumption for the data generating process in our context.

[25] The implicit assumption is, that the past affects the current state $r_t$ only through the most recent realization $r_{t-1}$. See Appendix A.8 for a summary of the main properties of Markov chains.

[26] This is consistent with the implications of some Walrasian batch type microstructure models, see e.g. Kyle (1985), Admati and Pfleiderer (1988), or Admati and Pfleiderer (1989).

Furthermore, the Markov switching model may be interpreted as a device to relax the somewhat restrictive assumption, that only one information event may occur per trading day, and that the information content is completely resolved into prices by the end of the same day. This assumption might be appropriate when news events consist e.g. of corporate announcements published after the close of the market, which are already known to a group of insiders earlier in the day. As noted by Hasbrouck (1999), it seems unrealistic from an empirical point of view to assume that all relevant private information comes exclusively in such a form. There are good reasons to allow for informational epochs with varying length of duration, which may either be very short termed epochs of only a few minutes stemming from transient market imbalances or uncertainty about the current composition of the trading population[27] or longer periods of time, e.g. when earnings announcements are known several days in advance of their public disclosure to a small group of insiders.

A more realistic model would therefore allow for information epochs, that may have irregular length, may begin at times other than the opening and end at times other than the close. Furthermore it could even be possible that different information epochs may overlap in time. However, given our limited knowledge of the real nature of relevant information events that occurred during the sample period for our selection of stocks, our model is designed to capture the aggregate effects of these phenomena. We allow for more than one information event per day, and information epochs that may have differential durations but are likely to have some impact persistence over the course of a trading day. These features may be captured by assuming that the regime variable $r_t$ follows an autoregressive process. The probabilistic structure of the sequential trade model with Markov switching regime probabilities is shown in Fig. 4.4.

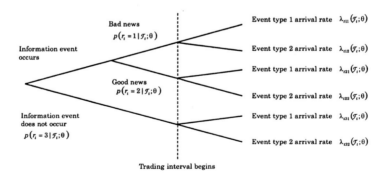

**Fig. 4.4.** Structure of the Markov switching sequential trade model.

---

[27] This resembles the heterogeneity model of Saar (2001a).

As in the case of the static mixture models, estimation of Markov switching regime models is most conveniently performed by using a variant of the EM-algorithm. The major difference between a static mixture and the Markov switching model is the implementation of the E-step. By analogy to the static mixture model, we define the variables $\xi_t^{(j)} = 1$ if $r_t = j$ and $\xi_t^{(ji)} = 1$ if $r_t = j$ and $r_{t-1} = i$ and zero otherwise. Conditioning on the initial states of the regime variable in $t = 1$, the *complete* log likelihood-function $\ln \mathcal{L}_C(\theta)$ is given by[28]

$$\ln \mathcal{L}_C(\theta) = \sum_{t=1}^{T} \sum_{j=1}^{N} \xi_t^{(j)} \cdot \ln f_j(y_t \mid r_t = j, \mathcal{F}_t; \theta) + \sum_{t=2}^{T} \sum_{j=1}^{N} \sum_{i=1}^{N} \xi_t^{(ji)} \cdot \ln p_{ji}, \quad (4.31)$$

where $\mathcal{F}_t$ again denotes the information set that is available at time $t$ and may include lagged dependent as well as contemporaneous and lagged exogenous variables, i.e. $\mathcal{F}_t = (y_1, \dots y_{t-1}, x_1, \dots x_t)$, and the parameter vector $\theta$ is given by[29]

$$\theta = (\beta_{11}, \dots, \beta_{NK}, p_{11}, \dots, p_{NN})',$$

when $f_j(y_t \mid r_t = j, \mathcal{F}_t; \theta)$ is assumed to be a Poisson density and

$$\theta = (\beta_{11}, \dots, \beta_{NK}, \alpha_{11}, \dots, \alpha_{NK}, p_{11}, \dots, p_{NN})',$$

when a density of the Negbin 2 type has been specified.

When evaluating the expectation of the complete data log-likelihood (4.31) in a Markov switching model, we have to draw probabilistic inferences on the unobservable quantities $\xi_t^{(j)}$. This task is accomplished by conducting a two step recursive algorithm, that will provide us with all the quantities necessary for estimation by the EM-algorithm.[30] In a first step, we therefore have to calculate *filtered* regime probabilities

$$\hat{\xi}_{t+1|t}^{(j)} \equiv p(r_{t+1} = j \mid \mathcal{Y}_t; \theta), \quad (4.32)$$

where $\mathcal{Y}_t = (y_t, \mathcal{F}_t)$ denotes the information set that includes all available information by time $t$, including contemporary values of the dependent variables

---

[28] If the initial state probabilities $p_j \equiv p(r_1 = j)$ are to be estimated along with the remaining parameters, the complete log-likelihood function is given by

$$\ln \mathcal{L}_C^*(\theta) = \ln \mathcal{L}_C(\theta) + \sum_{j=1}^{N} \xi_1^{(j)} \cdot \ln p_j.$$

[29] Again, only $N \times (N-1)$ transition probabilities may be treated as free parameters, since $p_{Ni} = 1 - \sum_{j=1}^{N-1} p_{ji}$. In order to keep the exposition as simple as possible, we will stick to the notation introduced above.

[30] A similar algorithm is known as 'forward-backward' algorithm in the statistics literature, see MacDonald and Zucchini (1997), pp. 92-93.

$y_t$. The filtered regime probability $\hat{\xi}_{t+1|t}^{(j)}$ may be interpreted as a one-step forecast of the conditional probability of observing regime $j$ at time $t+1$, given
the information set $\mathcal{Y}_t$ and can be calculated in a two-step recursion as follows

$$\hat{\xi}_{t|t}^{(j)} = \frac{\hat{\xi}_{t|t-1}^{(j)} \cdot f_j(y_t \mid r_t = j, \mathcal{F}_t; \theta)}{\sum\limits_{i=1}^{N} \hat{\xi}_{t|t-1}^{(i)} \cdot f_i(y_t \mid r_t = i, \mathcal{F}_t; \theta)} \qquad (4.33)$$

$$\hat{\xi}_{t+1|t}^{(j)} = \sum_{i=1}^{N} p_{ji} \cdot \hat{\xi}_{t|t}^{(i)}. \qquad (4.34)$$

The logic behind this recursion is quite easy to grasp. Equation (4.33) is
analogous to expression (4.12) in the static mixture model. The calculation
of $\hat{\xi}_{t+1|t}^{(j)}$ proceeds by employing the corresponding formula for the one step
forecast of a Markov chain.[31] With a given set of initial conditions[32] $\hat{\xi}_{1|0}^{(j)}$
and a given parameter vector, filtered regime probabilities can be calculated
iteratively.[33]

The probabilities $\hat{\xi}_{t|t-1}^{(j)}$ defined in (4.34) can be used to evaluate the in-
complete log-likelihood function $\ln \mathcal{L}_I(\theta)$.[34] This follows from noting that the
unconditional joint density is given by

$$f_t(y_t; \theta) = \sum_{j=1}^{N} \hat{\xi}_{t|t-1}^{(j)} \cdot f_j(y_t \mid r_t = j, \mathcal{F}_t; \theta). \qquad (4.35)$$

In the context of the EM-algorithm however these are inappropriate for the
evaluation of the expectation of the complete log-likelihood (4.31), since we
have to compute this expectation conditional on all observations in the sample
rather than using only the information available by time $t$. Therefore we use
*smoothed inferences* on the regime probabilities $\hat{\xi}_{t|T}^{(j)} \equiv p(r_t = j \mid \mathcal{Y}_T; \theta)$, which
may be calculated by means of a backward recursion, that starts from the fil-
tered inference $\hat{\xi}_{T|T}^{(j)}$ obtained from (4.33) and progresses iteratively according
to

$$\hat{\xi}_{t|T}^{(j)} = \hat{\xi}_{t|t}^{(j)} \cdot \sum_{i=1}^{N} \frac{p_{ij} \cdot \hat{\xi}_{t+1|T}^{(i)}}{\hat{\xi}_{t+1|t}^{(i)}}. \qquad (4.36)$$

---

[31] Thus, expression (4.34) is just a scalar notation of the forecast function for the
VAR representation of a homogenous Markov chain, see (A.30) in the Appendix.

[32] In our empirical application we will use the ergodic regime probabilities calculated
from (A.31) in the Appendix as initial conditions for the recursive computations
in (4.33) and (4.34).

[33] A static mixture model has time invariant forecasts of the regime probabilities
$\hat{\xi}_{t+1|t}^{(j)} = \pi_j$ for all $t$ but $\hat{\xi}_{t|t}^{(j)}$ is still varying with time.

[34] This would be the Markov switching model analog to (4.5).

This particular method of calculating $\hat{\xi}^{(j)}_{t|T}$ has been proposed by Kim (1994) and is valid only, when $r_t$ follows a first-order Markov chain and when the conditional density of $y_t$ depends only on the current state $r_t$ and on the filtration $\mathcal{F}_t$ which may consist of lagged dependent and exogenous variables that are independent of $r_t$.[35]

The E-step thus consists of taking the expectation of (4.31) conditional on the full sample of observable data $\mathcal{Y}_T = (y_1, \ldots, y_T, x_1, \ldots, x_T)$ and evaluating it using some preliminary guess for the parameter vector $\theta^{(p)}$. It can be shown[36], that the expectation of the complete log-likelihood $\ln \mathcal{L}_{EC}(\theta) \equiv E[\ln \mathcal{L}_{EC}(\theta) \mid \mathcal{Y}_T; \theta^{(p)}]$ is given by[37]

$$\ln \mathcal{L}_{EC}(\theta) = \sum_{t=1}^{T} \sum_{j=1}^{N} \hat{\xi}^{(j)}_{t|T} \cdot \ln f_j(y_t \mid r_t = j, \mathcal{F}_t; \theta^{(p)}) \tag{4.37}$$

$$+ \sum_{t=2}^{T} \sum_{j=1}^{N} \sum_{i=1}^{N} \hat{\xi}^{(ji)}_{t|T} \cdot \ln p_{ji}$$

where $\hat{\xi}^{(j)}_{t|T}$ is the full sample inference on the regime obtained by evaluating the backward recursion (4.36) using the preliminary guess $\theta^{(p)}$ and

$$\hat{\xi}^{(ji)}_{t|T} \equiv p\left(r_t = j \cap r_{t-1} = i \mid \mathcal{Y}_T; \theta^{(p)}\right) \tag{4.38}$$

$$= \hat{\xi}^{(i)}_{t-1|t-1} \cdot \frac{p_{ji} \cdot \hat{\xi}^{(j)}_{t|T}}{\hat{\xi}^{(j)}_{t|t-1}}.$$

The associated *M-step* consists of maximizing the expected complete log-likelihood function $\ln \mathcal{L}_{EC}(\theta)$ with respect to the parameter vector $\theta$. Again, application of the EM-algorithm to this estimation problem has the advantage, that the maximization of $\ln \mathcal{L}_{EC}(\theta)$ with respect to the parameters of the regression model and the transition probabilities can be conducted separately. The first order conditions lead to the following estimator for the transition probabilities[38]

$$\hat{p}_{ji} = \frac{\displaystyle\sum_{t=2}^{T} \hat{\xi}^{(ji)}_{t|T}}{\displaystyle\sum_{t=2}^{T} \hat{\xi}^{(i)}_{t-1|T}}, \tag{4.39}$$

---

[35] The rationale behind its calculation is given in Sect. A.9 in the Appendix.

[36] See Hamilton (1990).

[37] If the initial regime probabilities $p_j$ at $t = 1$ are to be included in the set of parameters to be estimated, the expected complete log-likelihood function is given by

$$\ln \mathcal{L}^*_{EC}(\theta) = \ln \mathcal{L}_{EC}(\theta) + \sum_{j=1}^{N} \hat{\xi}^{(j)}_{1|T} \cdot \ln p_j.$$

[38] See Appendix A.10.

which is essentially equal to the estimator for $p_{ji}$ that we would obtain if the regime variables $r_t$ were observable[39], with unobserved quantities replaced by appropriate probabilistic inferences.[40] The parameters of the regression model may be obtained by solving the $(K \times N)$ expressions

$$\frac{\partial \ell_{jk}}{\partial \theta} = \sum_{t=1}^{T} \hat{\xi}_{t|T}^{(j)} \cdot \left( \frac{\partial \ln f_{y_{tk}}(y_{tk} \mid r_t = j, \mathcal{F}_t; \theta)}{\partial \theta} \right) \stackrel{!}{=} 0. \qquad (4.40)$$

This is equivalent to the first order conditions of the static Poisson respectively Negbin 2 mixture regression models, with weights given by the smoothed regime probabilities $\hat{\xi}_{t|T}^{(j)}$. Therefore, after replacing $\hat{\xi}_t^{(j)}$ by $\hat{\xi}_{t|T}^{(j)}$, expressions (4.16) and (4.17) in the Poisson model, and (4.21) through (4.25) in the Negbin 2 model may be used to calculate the elements of the gradient vector and Hessian matrix that correspond to the regression and distributional parameters $\beta_{11}, \ldots, \beta_{NK}, \alpha_{11}, \ldots, \alpha_{NK}$. The gradient for the transition probabilities $p_{ji}$ is given by[41]

$$\frac{\partial \ln \mathcal{L}_{EC}(\theta)}{\partial p_{ji}} = \sum_{t=2}^{T} \left( \frac{1}{p_{ji}} \cdot \hat{\xi}_{t|T}^{(ji)} - \hat{\xi}_{t-1|T}^{(i)} \right), \qquad (4.41)$$

with second derivative equal to

$$\frac{\partial^2 \ln \mathcal{L}_{EC}(\theta)}{\partial p_{ji}^2} = -\sum_{t=2}^{T} \left( \frac{1}{p_{ji}^2} \cdot \hat{\xi}_{t|T}^{(ji)} \right). \qquad (4.42)$$

As in the case of the static mixture model, all cross derivatives of the kind $\frac{\partial^2 \ln \mathcal{L}_{EC}(\theta)}{\partial \beta_{jk} \partial p_{ji}}$, $\frac{\partial^2 \ln \mathcal{L}_{EC}(\theta)}{\partial \beta_{jk} \partial \beta'_{il}}$, $\frac{\partial^2 \ln \mathcal{L}_{EC}(\theta)}{\partial \beta_{jk} \partial \beta'_{jl}}$ and $\frac{\partial^2 \ln \mathcal{L}_{EC}(\theta)}{\partial p_{ji} \partial p_{kl}}$ for $i \neq j$ and $k \neq l$, as well as the corresponding cross derivatives involving $p_j$, are equal to zero. Expressions for the moments of the dependent variables can be derived from the unconditional joint density (4.35).[42]

---

[39] I.e. the frequency of observing a transition from state $i$ to state $j$ relative to the frequency of observing state $i$, see (A.33) in Appendix A.8.

[40] The corresponding estimator for the initial regime probabilities $p_j$ is $\hat{p}_j = \hat{\xi}_{1|T}^{(j)}$.

[41] The corresponding expressions for the initial probabilities are

$$\frac{\partial \ln \mathcal{L}_{EC}^*(\theta)}{\partial p_j} = \frac{1}{p_j} \cdot \hat{\xi}_{1|T}^{(j)} - 1,$$

and

$$\frac{\partial^2 \ln \mathcal{L}_{EC}^*(\theta)}{\partial p_j^2} = -\frac{1}{p_j^2} \cdot \hat{\xi}_{1|T}^{(j)}.$$

[42] See Appendix A.11 for a discussion.

## 4.2 Model Evaluation and Specification Testing

### 4.2.1 Specification Tests in Static Mixture and Markov Switching Models

There are two important issues in any mixture model, that have to be addressed in any serious empirical study:

1. Does the data come from a mixture distribution?
2. If the data comes from a mixture, how many support points (i.e. regimes) are needed to describe the data accurately?

If mixture models are used to describe real world data, there is always a problem of overfitting, since typically allowing more than one component will lead to a better fit of the model to the data, than the corresponding one component model. After all, the one component model is always nested in the mixture model.

On the other hand, statistical tests that could be used to address these questions are not easily derived because of the violation of some regularity conditions that must be satisfied when applying standard tests. Consider e.g. testing whether an univariate two component Poisson mixture model is more appropriate than a one component model, a hypothesis test of $H_0 : \lambda_1 = \lambda_2$ versus $H_A : \lambda_1 \neq \lambda_2$, could in principle be conducted by means of a likelihood ratio (LR) test, which is based on a comparison of the value of the likelihood function under the null and the alternative hypothesis.[43] However the corresponding test statistic will not have the usual $\chi^2$ distribution since the parameter $\pi_2$ is not identified under the null hypothesis. Therefore, the information matrix is singular under the null, and standard regularity conditions for an asymptotically valid test of $H_0$ fail to apply.[44] This problem is usually referred to as testing hypotheses when a *nuisance* parameter is not identified under the null hypothesis.

The asymptotic distribution of the LR-test statistic found by simulations differs essentially from the $\chi^2$. Furthermore, it appears to be model specific, in the sense, that the distribution is different if e.g. the marginal density of the observations is assumed to be of the Negbin 2 form instead of the

---

[43] To be more specific, the LR-test statistic is given by

$$\chi^2_{LR} = 2 \cdot [\ln \mathcal{L}_A(\hat{\theta}) - \ln \mathcal{L}_0(\hat{\theta})],$$

where $\mathcal{L}_0(\hat{\theta})$ is the value of the maximized likelihood function under the null hypothesis (i.e. assuming a one component model), and $\mathcal{L}_A(\hat{\theta})$ is the corresponding value under the alternative (i.e. assuming a two component mixture).

[44] See Hamilton (1990). Any application of the LR-test principle requires that the information matrix has full rank both under the null and the alternative hypothesis, see Spanos (1986), Chap. 16 or Davidson and MacKinnon (1993), Chap. 13 for a discussion.

Poisson.[45] Currently, to the best of our knowledge, no closed form distribution for the LR-test statistic has been established.[46] Another example is the usual $t$-statistic for $H_0 : \pi_j = 0$ against $H_A : \pi_j > 0$. Under the null, $\pi_j$ lies on the boundary of the admissible parameter space, thus violating one of the regularity conditions needed in order to derive the asymptotic normal distribution for the $t$-statistic.

Thus, when specification tests in static mixture and Markov switching models are being conducted, some care has to be exercised in order to avoid incorrect decisions as a result of the non-standard distributions of the test statistics involved. On the other hand, when the number of regimes is *known*, the maximum likelihood estimate of the parameter vector $\theta$ is consistent and has asymptotically a normal distribution with covariance matrix that may be derived from the usual estimates of the information matrix.[47] Hypothesis tests may be conducted in the usual fashion, as long as non of the maintained hypothesis violates the regularity conditions. Therefore, $t$-statistics for testing whether a particular regression parameter $\beta_{jk}$ is significantly different from zero may be compared to tabulated critical values of the $t$-distribution. Also, a test on whether different regimes occur with equal probability in the sample may be conducted by computing the usual $t$-statistic for $H_0 : \pi_j = \frac{1}{J}$ against $H_A : \pi_j \neq \frac{1}{J}$.

There are two basic approaches to circumvent the problems associated with statistical inference on the number of regimes in static mixture and Markov switching models. The first is to use test techniques especially developed for the situation, when nuisance parameters are present under the alternative hypothesis. A typical result in the literature dealing with this approach is that the asymptotic distribution of the test statistics are non-standard and will generally be model and data-dependent, so critical values of these tests cannot be tabulated.[48] Related tests that have been designed for use in mixture and Markov switching models are either based on bootstrapping[49] or other very demanding simulation techniques, which are prohibitively expensive in terms of computational effort when employed in the context of more complex models than the standard univariate Markov switching autoregressive model or when the sample size is large.[50]

Alternatively, one can estimate a $(N - 1)$-component mixture model and conduct a number of additional specification tests that avoid the pitfalls described above in order to see, whether an $N$-component model is needed to

---

[45] See Böhning, Dietz, Schaub, Schlattman, and Lindsay (1994).

[46] However, Aitken and Rubin (1985) employ a Bayesian estimation approach in the context of a static mixture model for which the LR-test statistic has an asymptotical $\chi^2$ distribution.

[47] See Lindgren (1978), Bickel and Ritov (1996), and Douc and Matias (2001).

[48] See Davies (1977), Davies (1987), Andrews and Ploberger (1994), and Hansen (1996b).

[49] See Feng and McCulloch (1996) and MacDonald and Zucchini (1997), pp. 96-97.

[50] See Hansen (1992), Hansen (1996a), Garcia (1998), and Coe (2002).

describe the data.[51] In the context of Markov switching regression models, Hamilton (1996) proposed a variety of such specification tests, which are based either on conditional moment conditions implied by the model or on the Lagrange multiplier (LM) principle. Both types of tests require the evaluation of the score function $s_t$, which is defined as

$$s_t = \frac{\partial \ln f_t(y_t; \theta)}{\partial \theta}. \tag{4.43}$$

Hamilton derives analytical expressions for the scores of the incomplete log-likelihood function $\ln \mathcal{L}_I(\theta)$, leading to different formulas for $s_t$ than those given in Sect. 4.1, which were derived from the expected complete log-likelihood $\ln \mathcal{L}_{EC}(\theta)$. In general, the two score functions will not be identical. However, as noted by Ruud (1991), maximization of $\ln \mathcal{L}_{EC}(\theta)$ will yield the same parameter estimates as maximization of $\ln \mathcal{L}_I(\theta)$, so the expected scores of both functions will be equal to zero when evaluated at the parameter estimates.[52] For this reason, when the two score functions are being evaluated using the parameter estimates $\hat{\theta}$ found by maximizing either of the two likelihood functions, they are equivalent to each other. Hence, all of the specification tests proposed by Hamilton can in principle be conducted as well in our framework using the scores of the expected complete log likelihood function. In the following Sections we will describe methods, that will be used to determine the appropriateness of the static mixture and Markov switching approach in our empirical applications.

### 4.2.2 Determining the Number of Regimes

Diagnostic tools that may be used to assess the appropriateness of a mixture model have been introduced by Lindsay and Roeder (1992). Their approach is based on the *residual function* defined as

$$c(y, \hat{\theta}) = \frac{\tilde{f}(y)}{f(y \mid \hat{\theta})} - 1, \tag{4.44}$$

where

$$\tilde{f}(y) = T^{-1} \cdot \sum_{t=1}^{T} I(y_t = y)$$

is the observed relative frequency of $y$ in the sample, $I(\cdot)$ is the indicator function taking the value 1 when $y_t = y$ and zero otherwise, and

$$f(y \mid \hat{\theta}) = T^{-1} \cdot \sum_{t=1}^{T} f_t(y \mid \hat{\theta})$$

---

[51] See Hamilton (1993).
[52] See also Appendix A.3.

is the expected frequency of observing the value $y$ given the estimated parameter vector $\hat{\theta}$. If the estimated model is based on a one-component density, the shape of the plot of the resulting *homogeneity residual* against $y$ indicates whether a mixture model is appropriate for the given data set or not. A data generating process that is a mixture of densities belonging to the one parameter exponential family will have a residual plot with convex shape and two sign changes $(+, -, +)$. Intuitively, this reflects that a $N$-component mixture density contains more probability mass in its tails, than the corresponding one component density.[53] The Poisson distribution belongs to the class of the one parameter exponential family, while the negative binomial does not. In spite of this, Trivedi and Deb (1997) advocate the use of residual plots for a negative binomial mixture model as well on a heuristic basis.

Evaluating $c(y, \hat{\theta})$, using maximum likelihood estimates for a $N$-mixture model for the quantity in the denominator, may be indicative of whether a $(N + 1)$-mixture is needed to fit the data appropriately. A plot of the *mixture residuals* that is close to a horizontal line through 0 supports the more parsimonious $N$-component model, with the homogeneity model as a special case when $N = 1$. Pointwise confidence intervals for the residual plots may be constructed by noting that $c(y, \hat{\theta})$ is approximately normally distributed with variance

$$
Var\left[c(y, \hat{\theta})\right] = \frac{1}{T} \cdot \left( \frac{1 - f(y \mid \hat{\theta})}{f(y \mid \hat{\theta})} - \frac{[y - T^{-1} \cdot \sum_{t=1}^{T} E(y_t \mid \hat{\theta})]^2}{T^{-1} \cdot \sum_{t=1}^{T} Var(y_t \mid \hat{\theta})} \right), \quad (4.45)
$$

where $E(y_t \mid \hat{\theta})$ is the conditional mean of $y_t$ under the estimated model and $Var(y_t \mid \hat{\theta})$ is the corresponding conditional variance. Note, that the behavior of the residual function becomes very unstable when the range of $y_t$ is unbounded and the function is being evaluated in regions where sample observations are sparse or where the probabilities $f(y \mid \hat{\theta})$ are small. In the case of the Poisson density $f(y \mid \hat{\theta})$ goes rather quickly to zero, as we move away from the mean of the distribution. Therefore it is recommended to compute a truncated version of the residual function, restricting the range of $y$-values at which $c(y, \hat{\theta})$ is being evaluated to an appropriate subset of the observed range of sample values.

Leroux (1992) shows that under some regularity conditions the number of components can be estimated consistently using information criteria. The results of Monte Carlo experiments reported in Trivedi and Deb (1997) and in Wang, Cockburn, and Puterman (1998) provide evidence for the practical success of information criteria in discriminating between alternative data generating processes. In our empirical applications, we will therefore employ the Akaike information criterion ($AIC$) and the Bayesian information criterion

---

[53] See Shaked (1980) for a proof.

$(BIC)$.[54] These are given by

$$AIC = -2\ln \mathcal{L}(\theta) + 2 \cdot q \qquad (4.46)$$

$$BIC = -2\ln \mathcal{L}(\theta) + (\ln T) \cdot q \qquad (4.47)$$

where $\mathcal{L}(\theta)$ denotes the value of the observed (incomplete) likelihood function, $q$ denotes the number of estimated parameters, and $T$ is the number of observations. The criteria $AIC$ and $BIC$ give increasingly large penalties in $T$ and $q$. The preferred model is therefore associated with the minimal information criterion. It is well known, that the $AIC$ often tends to overestimate the true dimension of a model, while it never underestimates the dimension asymptotically, thus suggesting a superiority of the $BIC$.[55]

### 4.2.3 A Conditional Moment Test for Goodness of Fit

One way to assess the goodness of fit of the models proposed in this study is to compare the corresponding fitted densities for the dependent count variables with actually observed frequencies. A formal test can be conducted by using a generalization of the $\chi^2$ goodness of fit test.[56] The null hypothesis is that the density of the vector of dependent variables $y_t$ is given by (4.4), or (4.35) when a Markov switching model is specified, with conditional densities $f_{y_{tk}}(y_{tk} \mid r_t = j; \theta)$ replaced either by the Poisson or the Negbin specification, whatever is appropriate. Thus the null hypothesis may be stated as follows

$$H_0: \quad E[\tilde{F}_t(y) - F_t(y, \hat{\theta})] = 0,$$

against

$$H_A: \quad E[\tilde{F}_t(y) - F_t(y, \hat{\theta})] \neq 0,$$

where $\tilde{F}_t(y)$ is the $(C \times 1)$ vector of indicator functions $I(y_t \in c_1), \ldots, I(y_t \in c_C)$, which comprise the observed distribution of the dependent count variables $y_t$, $C$ is the prespecified number of cells employed for fitting the distributions. The set of cells $c_1, \ldots, c_C$ constitutes a partition of the support of the dependent variables. $F_t(y, \hat{\theta})$ is the corresponding vector of fitted cell probabilities

$$\sum_{y \in c_1} f_t(y; \hat{\theta}), \ldots, \sum_{y \in c_C} f_t(y; \hat{\theta})$$

with $f_t(\cdot; \hat{\theta})$ as defined in (4.4) or (4.35) and $\hat{\theta}$ is the maximum likelihood estimate of $\theta$. The corresponding test statistic is

$$\chi^2_{CM} = \sqrt{T} \cdot m(\hat{\theta})' \cdot \hat{V}_m^{-1} \cdot m(\hat{\theta}), \qquad (4.48)$$

---

[54] See Akaike (1973) and Schwarz (1978).

[55] See Sawa (1978) for a discussion.

[56] This form of the test was originally suggested by Heckman (1984). See Newey (1985), Tauchen (1985), Andrews (1988a) and Andrews (1988b) for related work.

where the sample moments $m(\hat{\theta})$ are given by

$$m(\hat{\theta}) = \frac{1}{T} \cdot \sum_{t=1}^{T} m_t = \frac{1}{T} \cdot \sum_{t=1}^{T} \left[ \tilde{F}_t(y) - F_t(y, \hat{\theta}) \right], \qquad (4.49)$$

$\hat{\theta}$ is the estimated vector of parameters, $\hat{V}_m$ is a consistent estimate of the asymptotic covariance matrix of the sample moments $m(\hat{\theta})$ and $\hat{V}_m^{-1}$ denotes the inverse of $\hat{V}_m$. The null hypothesis is rejected, if $\chi^2_{CM}$ is sufficiently large. The asymptotic distribution of the test statistic under the null is a chi-square distribution with degrees of freedom given by the rank of the covariance matrix $V_m$, which is equal to the number of cells minus one in our application.

If the parameter vector $\theta$ is estimated efficiently, a simplified way of computing the test statistic $\chi^2_{CM}$ may be employed.[57] This is achieved by computing $T$ times the uncentered $R^2$ from an auxiliary regression of the form

$$1 = s'_t \cdot \beta_s + m'_t \cdot \beta_m + \text{residual},$$

where the $(q \times 1)$ vector of scores $s_t$ as defined in (4.43) and the $(C - 1 \times 1)$ vector of moment conditions[58] $m_t$ are used as regressors. Thus, the test statistic is given by

$$\chi^2_{CM} = \left[ \sum_{t=1}^{T} s'_t \quad \sum_{t=1}^{T} m'_t \right] \cdot \left[ \begin{matrix} \sum\limits_{t=1}^{T} s_t s'_t & \sum\limits_{t=1}^{T} s_t m'_t \\ \sum\limits_{t=1}^{T} m_t s'_t & \sum\limits_{t=1}^{T} m_t m'_t \end{matrix} \right]^{-1} \cdot \left[ \sum_{t=1}^{T} s'_t \quad \sum_{t=1}^{T} m'_t \right]'. \quad (4.50)$$

In our empirical applications we will evaluate $m_t$ over the set of cells $c_1 : y_t = 0, c_2 : y_t = 1, \ldots, c_C : y_t \geq y^*$ where $y^*$ denotes a predefined threshold value defining the last cell. In multivariate applications we will evaluate $\chi^2_{CM}$ for each variable separately, thus concentrating on the marginal rather than on the joint distribution. In the calculation of $\chi^2_{CM}$ according to (4.50), it is always the contribution of the last cell $c_C$ that is being omitted.

### 4.2.4 Testing Parameter Restrictions

In order to conduct specification tests in our framework that may be used to compare the generalized sequential trade model variants implied by our econometric models to the original EKOP model, we would want to test for the validity of several restrictions on the estimates of the regression parameters across different regimes and equations. Suppose we have estimated a static mixture or Markov regression switching model by maximum likelihood and denote the estimate of the parameter vector by $\hat{\theta}$.

---

[57] See Andrews (1988a).

[58] One cell is being dropped in order to circumvent multicollinearity problems as a consequence of the $C$ cell probabilities summing to one.

Let us partition the $(q \times 1)$ parameter vector $\hat{\theta}$ as $\hat{\theta} = [\beta_1, \beta_2, \beta_3, \beta_4, \theta_0]'$, where $\beta_1$ through $\beta_4$ denote four regression parameters and $\theta_0$ has dimension $(q_0 \times 1)$ and contains the remaining parameter estimates for which we do not want to impose any restrictions. We want to test the joint hypothesis whether $\beta_1 = \beta_2$ and $\beta_3 = \beta_4$, so the null hypothesis may be stated as $H_0 : \beta_1 - \beta_2 = 0 \cap \beta_3 - \beta_4 = 0$ and the alternative is $H_A : \beta_1 - \beta_2 \neq 0 \cup \beta_3 - \beta_4 \neq 0$. In order to conduct this specification test we may apply a Wald test, whose test statistic is of the following general form[59]

$$\chi^2_W = T \cdot G(\hat{\theta})' \cdot \left\{ \left( \frac{\partial G(\hat{\theta})}{\partial \hat{\theta}} \right)' \cdot I(\hat{\theta})^{-1} \cdot \left( \frac{\partial G(\hat{\theta})}{\partial \hat{\theta}} \right) \right\}^{-1} \cdot G(\hat{\theta}), \qquad (4.51)$$

where $G(\hat{\theta})$ denotes a $(g \times q)$ matrix representing $g$ possibly nonlinear restrictions of the form $G(\theta) = 0$, $\frac{\partial G(\hat{\theta})}{\partial \hat{\theta}}$ denotes the $(g \times q)$ matrix of partial derivatives of the restrictions with respect to the elements of $\hat{\theta}$, $T$ is the sample size, and $I(\hat{\theta})$ denotes any consistent estimate of the information matrix evaluated using the unrestricted ML-estimates of the parameter vector. The asymptotic distribution of the Wald test statistic is the $\chi^2$ with $g$ degrees of freedom under the null hypothesis. Thus, for large values of $\chi^2_W$ the null hypothesis is rejected.

In our empirical applications we will want to test linear equality restrictions among sets of regression parameters in the first place. Thus the restrictions will always be of the form $G(\hat{\theta}) = G \cdot \hat{\theta}$, with $G$ an $(g \times q)$ matrix that describes the equality constraints we are interested in. In our initial example $G$ is a $(2 \times q)$ matrix given by

$$G = \begin{bmatrix} 1 & -1 & 0 & 0 & o_{q_0} \\ 0 & 0 & 1 & -1 & o_{q_0} \end{bmatrix},$$

where $o_{q_0}$ denotes a $(1 \times q_0)$ vector of zeros. In the case of linear restrictions, the matrix of partial derivatives takes on the convenient form

$$\frac{\partial G(\hat{\theta})}{\partial \hat{\theta}} = G',$$

and the Wald test statistic may be computed as

$$\chi^2_W = T \cdot \hat{\theta}' \cdot G' \cdot \left\{ G \cdot I(\hat{\theta})^{-1} \cdot G' \right\}^{-1} \cdot G \cdot \hat{\theta}. \qquad (4.52)$$

Employing a Wald test in the context of our framework has the great advantage that only estimates of the unrestricted model are needed, while the corresponding likelihood ratio and Lagrange multiplier tests require estimation of the restricted model, which in our case is very difficult to conduct, as it

---

[59] See Spanos (1986), p. 333.

would force us to maximize the incomplete likelihood function $\mathcal{L}_I(\theta)$ directly. However, $\mathcal{L}_I(\theta)$ may be expected to be very ill-behaved in practice, since one has to take into account restrictions across equations as well as across different regimes. In this case we would be unable to exploit the convenience of using the EM-algorithm. Note furthermore, that the Wald tests we are going to employ do not suffer from the problems mentioned in Sect. 4.2.1, since none of the hypothesis we will specify involves any unidentified nuisance parameters under the null hypothesis and none of the parameters lies on the boundary of its domain. We therefore expect, that the test statistic should have asymptotically a $\chi^2$ distribution with $g$ degrees of freedom as usual.

### 4.2.5 Testing for Autocorrelation

Since we intend to estimate dynamic models for discrete valued random variables, it is important to assess whether the specified model does indeed capture the dynamics of the dependent variable accurately. Often, such tests are based on whether the regression residuals are white noise processes, i.e. whether they have constant, finite variance, and are serially uncorrelated. However, the count data models we employ are regression models for non-negative, heteroscedastic random variables. Therefore, the raw residuals $\tilde{u}_{tk} = y_{tk} - \hat{y}_{tk}$ will always appear to be non-stationary, due to the fact, that the conditional variance of $y_{tk}$ is a function of the conditional mean and varies in time.[60]

In order to apply standard tests for residual autocorrelation, the raw residuals have to be standardized in some way. It is customary to employ the Pearson residuals

$$\hat{u}_{tk} = \frac{y_{tk} - \hat{y}_{tk}}{\sqrt{\hat{v}_{tk}}}, \tag{4.53}$$

where $\hat{v}_{tk}$ denotes the estimated conditional variance of $y_{tk}$, for tests of autocorrelation. Note that the forecast $\hat{y}_{tk} \equiv E(y_{tk} \mid \mathcal{F}_t; \hat{\theta})$ and conditional variance function $\hat{v}_{tk} \equiv Var(y_{tk} \mid \mathcal{F}_t; \hat{\theta})$ are given by

$$\hat{y}_{tk} = \sum_{j=1}^{N} \hat{\pi}_j \cdot \hat{\lambda}_{tjk} \cdot \Delta t$$

$$\hat{v}_{tk} = \sum_{j=1}^{N} \hat{\pi}_j \cdot \hat{v}_{tjk} + \sum_{j=1}^{N} \hat{\pi}_j \cdot (\hat{\lambda}_{tjk} \cdot \Delta t - \hat{y}_{tk})^2,$$

in static mixture models, and

$$\hat{y}_{tk} = \sum_{j=1}^{N} \hat{\xi}_{t|t-1}^{(j)} \cdot \hat{\lambda}_{tjk} \cdot \Delta t$$

---

[60] See Cameron and Trivedi (1998), p. 228.

$$\hat{v}_{tk} = \sum_{j=1}^{N} \hat{\xi}_{t|t-1}^{(j)} \cdot \hat{v}_{tjk} + \sum_{j=1}^{N} \hat{\xi}_{t|t-1}^{(j)} \cdot (\hat{\lambda}_{tjk} \cdot \Delta t - \hat{y}_{tk})^2,$$

in Markov switching models, with regime specific conditional variance function $\hat{v}_{tjk} = \hat{\lambda}_{tjk} \cdot \Delta t$, in the Poisson model, and $\hat{v}_{tjk} = \hat{\lambda}_{tjk} \cdot \Delta t + \hat{\alpha}_{jk} \cdot (\hat{\lambda}_{tjk} \cdot \Delta t)^2$, when a Negbin 2 model has been specified.[61]

If the model is correctly specified, then $\hat{u}_{tk}$ should have a constant variance and be uncorrelated with $\hat{u}_{sk}$, for $s \neq t$. A first test of whether the dynamics of $y_t$ are correctly captured by the estimated model, is to plot the autocorrelation function (ACF) of the Pearson residuals against the lag length. An estimate of the $m$-th order autocorrelation $\hat{\rho}_m$ is given by

$$\hat{\rho}_m = \frac{\sum_{t=m+1}^{T} \hat{u}_{tk} \cdot \hat{u}_{(t-m)k}}{\sum_{t=1}^{T} \hat{u}_{tk}^2}, \tag{4.54}$$

and is asymptotically normally distributed with zero mean and variance $T^{-1}$ under the null hypothesis of no autocorrelation. A test on whether a single autocorrelation coefficient $\rho_m$ is equal to zero can be based on comparing $\sqrt{T} \cdot \hat{\rho}_m$ to tabulated quantiles of the standard normal distribution.[62]

Formal tests of whether all autocorrelations up to order $m$ are equal to zero are based on the estimates of the ACF. The test statistic

$$\chi_{BP}^2 = T \cdot \sum_{s=1}^{m} \hat{\rho}_s^2, \tag{4.55}$$

has been suggested by Box and Pierce (1970) and follows asymptotically a $\chi^2$ distribution with $m$ degrees of freedom. A refined version of this test with better small sample performance has been suggested by Ljung and Box (1978). The Ljung-Box test statistic is given by

$$\chi_{LB}^2 = T \cdot (T+2) \cdot \sum_{s=1}^{m} (T-s)^{-1} \cdot \hat{\rho}_s^2, \tag{4.56}$$

and has a $\chi^2$ distribution with $m$ degrees of freedom as well. Thus, for large values of the test statistics $\chi_{BP}^2$, respectively $\chi_{LB}^2$, the Null hypothesis of no residual autocorrelation up to lag $m$ will be rejected.

---

[61] See Appendix A.6 and A.11.

[62] Note that the asymptotic normality of the sample autocorrelations $\hat{\rho}_m$ does not require that the stochastic process $u_t$ has normal distribution as well, but only that its variance is finite, see Anderson and Walker (1964).

## 4.3 Mixture and Regime Switching Models in Econometrics

The econometric models we proposed in this Chapter for the estimation of sequential trade models based on intradaily time series for event counts all belong to the class of static mixture and Markov switching models, that have a rich history in the statistical and econometric literature.[63]

Mixture models have often been employed in the context of event history analysis in order to model unobserved individual heterogenity in cross-sectional or panel data.[64] There are also many applications of univariate static mixture models for count data, e.g. to assess the effects of direct marketing on purchases (Wedel, Desarbo, Bult, and Ramaswamy (1993)) and treatment effects in clinical trials (Wang, Puterman, Cockburn, and Le (1996)), as a model of the demand for health care (Trivedi and Deb (1997), Deb and Trivedi (2002)), to analyze the relationship between patents and research and development spending of individual firms (Wang, Cockburn, and Puterman (1998)), or as a model for criminal careers in sociological studies[65]. However, none of these papers treats the case of a time series of events. A notable exception is the material summarized in the book by MacDonald and Zucchini (1997), who motivate hidden Markov models as an alternative to other time series models for discrete valued processes, but they do not treat autoregressive specifications of the conditional mean of the observable time series.

Regression models that allow for regime dependent conditional mean specifications have a long history in econometrics. Apart from the literature on testing for structural breaks[66], *switching regression* models that allow for repeated discrete changes of regime have been in use mainly as models for macroeconomic time series with different behavior in recessions and in expansion phases. In these models, changes in the regime are modelled as the outcome of an unobserved, discrete random variable which identifies the state of the economy in each period. Extensions of this approach lead to models where the regime variable is itself an autoregressive process, whose behavior is governed by a hidden Markov chain.[67]

In a seminal paper Hamilton (1989) combined the Markov chain approach for the latent regime with autoregressive dynamics in the observed economic

---

[63] Surveys on static mixture and Markov switching models are given in Titterington, Smith, and Makov (1985), Hamilton (1993), Hamilton (1994), Chap. 22, Lindsay and Lesperance (1995), Kim and Nelson (1999), Raj (2002), and Hamilton and Raj (2002).

[64] See e.g. Heckman and Singer (1982), Gritz (1993), Heckman and Taber (1994), Allenby, Leone, and Jen (1999) and Land, Nagin, and McCall (2001).

[65] See e.g. Nagin and Land (1993), D'Unger, Land, McCall, and Nagin (1998), Roeder, Lynch, and Nagin (1999), and D'Unger, Land, and McCall (2002)

[66] See e.g. Chow (1960), Goldfeldt and Quandt (1965), or Perron (1989).

[67] See e.g. Quandt (1958), Quandt (1972), Goldfeldt and Quandt (1973), Quandt and Ramsey (1978), Richard (1980), and Cosslett and Lee (1985).

time series. His Markov switching autoregressive model (MSAR) has often been used to model macroeconomic and financial time series. Typical applications of his approach involve the analysis of business cycles (Lam (1990), Hamilton and Perez-Quiros (1996)), the determinants of exchange rate fluctuations (Engel and Hamilton (1990), Mundaca (2000), and Dewachter (2001)) and interest rate movements (Garcia and Perron (1996), Smith (2002), and Ang and Bekaert (2002)), the analysis of asset pricing models (Turner, Startz, and Nelson (1989), Bekaert and Harvey (1995), and Schaller and van Norden (2002)) and the determinants of voting behavior (Blomberg (2000)).

The MSAR model has been extended in many ways, e.g. by allowing for time-varying transition probabilities in the Markov chain (Filardo (1994), Diebold, Lee, and Weinbach (1997)), to model changes in the conditional variances in GARCH models (Cai (1994), Hamilton and Susmel (1994), and Gray (1996)), in connection with dynamic factor models (Diebold and Rudebusch (1996) and Chauvet, Juhn, and Potter (2002)), and for the analysis of cointegration relations in vector autoregressive models (Krolzig, Marcellino, and Mizon (2002)). Extending the autoregressive conditional duration model introduced by Engle and Russell (1998), Hujer, Vuletić, and Kokot (2002) recently introduced the class of Markov switching ACD model which can be used to forecast time series of durations between successive trade events.

The common link between the MSAR model and the literature on switching and mixture regression models is that both approaches imply that the data generating process of the dependent variable can be described by a discrete mixture density, where the conditional density of the dependent variable, given the regime, is specified to be from some known family of distributions, usually the Gaussian, and the density of the regime variable is left unspecified. The regime probabilities are estimated non-parametrically along with the regression parameters, by imposing the restriction, that the regime density is discrete valued and has a finite number of support points.[68] It is possible to derive moment-based estimators for mixture regression models, although the evidence from simulation studies, that compare general method of moments (GMM) and ML estimators does not encourage the use of GMM estimators at all.[69] There are also many applications that employ Bayesian estimation techniques to mixture and MSAR models.[70]

---

[68] The assumption that the regime variable is discrete valued is not necessary. As shown by Laird (1978) and Jewell (1982), the ML-estimator in the case of a continuous valued, unobservable regime variable with unknown density also has at most a finite number of support points, and can therefore be estimated consistently by the same methods that we employed in the discrete valued case.

[69] See Heckman, Robb, and Walker (1990) and Deb, Ming, and Trivedi (1999).

[70] See Hamilton (1991), Diebolt and Robert (1994), and McCulloch and Tsay (1994).

# 5

# Empirical Results

## 5.1 The TAQ Database

The Trade and Quote (TAQ) database is a collection of intraday trades and quotes for all securities listed on the New York Stock Exchange, American Stock Exchange, NASDAQ and on associated regional exchanges (Boston, Cincinnati, Midwest, Pacific, Philadelphia) and is available at relatively low cost in electronic form on a monthly basis from the New York Stock Exchange Inc. The TAQ data set contains information about the timing of the trades, transaction prices and volumes as well as every revision of best bid and ask prices and corresponding volumes. Furthermore, the TAQ data base contains information related to the conditions under which the trade took place (e.g. whether an order was stopped by the market maker, or whether the transaction calls for delivery and payment on the same trading instead of the usual three day limit) and a correction indicator, that identifies trades that were incorrectly recorded and had to be corrected later on.[1]

The data comes in two files, the trade data set, containing time-stamped information on all transactions that have been conducted on the participating stock exchanges and the quote data set, that contains time-stamped information on all associated quote revisions. In the first step of our data preparation we extracted all trade and quote information for the common shares of five stocks, Boeing (BA), Disney (DIS), IBM, Coca-Cola (KO) and Exxon (XON). We use all non-erroneous transactions and quote revisions that occurred on the NYSE during the regular trading day (9.30 a.m. until 4.00 p.m.). Our sampling period spans a complete month (22 trading days), from August 1 until August 31, 1996. Since the trading times have been recorded with a precision measured in seconds, we aggregate consecutive trades that occur within the same second, by summing the corresponding volumes and computing a volume weighted average of their transactions prices. Trades occurring within

---

[1] See Hasbrouk and Sofianos (1993), Hasbrouck, Sofianos, and Sosebee (1993), or TAQ (1996) for further details.

one second are very likely to result from batched orders, i.e. transactions with more than one participant on either side of the trade. These trades constitute de facto one huge transaction, but since there is no information in the original data set that would allow us to identify trades resulting from batched orders, we proceed by consolidating the trades as stated above.

Table 5.1 contains some summary statistics for the TAQ data set. It is worthwhile noting that all the variables we examine are serially dependent, with Ljung-Box statistics that are far from their critical values at conventional significance levels. The column entitled 'Duration' contains summary statistics for the duration between successive trades measured in seconds.[2] This variable is of special interest in the context of our study, since it is closely related to the trading intensity that we aim to explain using count data models. As laid out in Sect. 3.6, trade durations can be used as an alternative empirical basis for the estimation of sequential trade models.[3] If the arrival rates of informed and uninformed traders are indeed governed by independent Poisson processes as in the EKOP model, we would expect that the durations between successive buys and sells of informed and uninformed traders are exponentially distributed, conditional on the information regime.[4] Therefore the observed trade durations should follow a mixture of exponential distributions, and thus be overdispersed relative to a one-component exponential distribution.[5] This is indeed the case for all stocks in our sample, but this feature by itself should not be interpreted as sufficient evidence in favor of the EKOP model, because the apparent overdispersion might be explained by other models as well, e.g. the durations might follow a Weibull distribution instead of an exponential.

Although all the stocks in our sample are blue chips and thus belong to the most frequently traded assets on the NYSE, there are significant differences in the frequency of trading during our sample period. The number of transactions ranges from 7583 trades (BA) to 15445 trades for the most frequently traded share in our sample (IBM). Note that the average trade size is however roughly the same order of magnitude for both stocks, so is the variation in the volume of the trades. In our sample there are stocks with both larger average volume and larger variation in volume (KO, XON) as well as smaller average volume (DIS), so the volume per trade does not seem to be related to the frequency of trading in any obvious manner.

Another interesting feature is the size of the bid-ask spread. In general, the size of the spread appears to be very low for our sample of stocks, with average size between 0.143 US-$ (KO) and 0.184 US-$ (BA), which is well below the mark of two ticks (0.25 US-$). All of the stocks have a median spread equal to one tick (0.125 US-$), which is by far the most often observed value in our

---

[2] Censored spells occurring at the beginning and the end of each trading day have been dropped before the sample statistics of the trade durations have been calculated.

[3] See Hujer, Vuletić, and Kokot (2002) for details.

[4] See Appendix A.1 for a derivation of this result.

[5] The exponential distribution has standard deviation equal to its mean.

**Table 5.1.** Summary statistics for the TAQ data

| Statistic | Variable | | | |
|---|---|---|---|---|
| | Duration | Volume | Spread | P.-spread |
| **Boeing** | | | | |
| Obs. | 7583 | 7583 | 6178 | 6178 |
| Mean | 67.66 | 2313.00 | 0.184 | 0.20 |
| Std.-error | 89.74 | 6163.20 | 0.064 | 0.07 |
| Minimum | 1 | 100 | 0.125 | 0.13 |
| Median | 36 | 800 | 0.125 | 0.14 |
| Maximum | 1021 | 221900 | 0.500 | 0.56 |
| $\rho$ | 0.07 | 0.01 | 0.04 | 0.04 |
| $\chi^2_{LB}$ | 2732.79 | 320.92 | 1538.60 | 1580.03 |
| **Disney** | | | | |
| Obs. | 9659 | 9659 | 10190 | 10190 |
| Mean | 53.23 | 1841.95 | 0.152 | 0.26 |
| Std.-error | 68.31 | 4932.31 | 0.054 | 0.09 |
| Minimum | 1 | 100 | 0.125 | 0.21 |
| Median | 30 | 500 | 0.125 | 0.22 |
| Maximum | 1198 | 141700 | 0.500 | 0.89 |
| $\rho$ | 0.04 | 0.04 | 0.03 | 0.03 |
| $\chi^2_{LB}$ | 1535.67 | 170.33 | 3145.01 | 3262.62 |
| **IBM** | | | | |
| Obs. | 15445 | 15445 | 9151 | 9151 |
| Mean | 33.20 | 2499.22 | 0.170 | 0.15 |
| Std.-error | 48.17 | 6636.51 | 0.064 | 0.06 |
| Minimum | 1 | 100 | 0.125 | 0.11 |
| Median | 17 | 1000 | 0.125 | 0.11 |
| Maximum | 967 | 300000 | 0.500 | 0.46 |
| $\rho$ | 0.11 | 0.00 | 0.01 | 0.01 |
| $\chi^2_{LB}$ | 15197.99 | 312.68 | 827.28 | 849.29 |
| **Coca-Cola** | | | | |
| Obs. | 11985 | 11985 | 7119 | 7119 |
| Mean | 42.81 | 3127.85 | 0.143 | 0.28 |
| Std.-error | 53.41 | 9403.80 | 0.044 | 0.09 |
| Minimum | 1 | 100 | 0.125 | 0.24 |
| Median | 25 | 500 | 0.125 | 0.25 |
| Maximum | 705 | 528800 | 0.375 | 0.76 |
| $\rho$ | 0.04 | 0.00 | 0.00 | 0.00 |
| $\chi^2_{LB}$ | 1494.21 | 236.76 | 282.34 | 270.25 |
| **Exxon** | | | | |
| Obs. | 9467 | 9467 | 9731 | 9731 |
| Mean | 53.84 | 3092.72 | 0.155 | 0.19 |
| Std.-error | 72.16 | 7147.99 | 0.054 | 0.07 |
| Minimum | 1 | 100 | 0.125 | 0.15 |
| Median | 29 | 1000 | 0.125 | 0.15 |
| Maximum | 1597 | 246800 | 0.500 | 0.60 |
| $\rho$ | 0.05 | -0.01 | 0.01 | 0.01 |
| $\chi^2_{LB}$ | 2048.03 | 126.18 | 1305.78 | 1367.11 |

Obs. is the number of observations, P.-spread is the spread relative to the bid-ask midpoint (in 100%), $\rho$ is the estimated autocorrelation coefficient for 50 lags, $\chi^2_{LB}$ is the Ljung-Box statistic for 50 lags. The corresponding tabulated 99% (95%) critical value for 50 degrees of freedom is 76.2 (67.5).

sample. Moreover, the spread is never any higher than four ticks (0.5 US-$) in our sample. Most of the observations in the quote data set appear to be related to changes of the market depth (i.e. the volume associated with the best bid resp. ask quote) rather than to changes of the spread. Note that while the spread expressed in US-\$ appears to be smaller for more frequently traded stocks, we find no such tendency if we relate the spread to the bid-ask midpoint. The average percentage spread is highest for KO which is at the same time the second most often traded stock in our sample, while it is lowest for IBM, the most frequently traded stock.

## 5.2 The Trade Direction

### 5.2.1 Algorithms for the Determination of the Trade Direction

An important piece of information for conducting empirical studies related to market microstructure issues, is the *trade direction*, i.e. whether a trade was initiated by the buyer or the seller. If the trading mechanism involves the presence of a single, delegated market maker who participates in all trades, the trade direction could be easily determined as buyer-initiated, if the market maker sells shares to other traders and as seller initiated if he buys shares from other traders. The main drawback of this method is of course, that it demands information on the counter parties of each trade that is rarely included in high-frequency data sets available to researchers. Since most of the available high frequency data sets, including the TAQ data set do not contain explicit information about the participants of the trade, empirical studies typically involve the construction of an indicator variable identifying the initiator of each trade from the data at hand.

If at least information on the prevailing bid-ask spread at the time of the trade is available, any trade occurring at the ask could be classified as a buy, while trades at the bid would be sells. In many situations however, this simple classification scheme will not work, either because the quote information is insufficient or not included at all. Even when quotes are available, the simple quote matching scheme will often not work for all trades, since typically a significant fraction of the trades occurs neither at the ask nor at the bid. In these situations improved classification schemes can be applied, which incorporate trading regulations and customs known to prevail at the stock exchange where the data was collected. Some generally applicable classification techniques are given in the following:[6]

- *Tick test*: The current trade is classified as buyer (seller)-initiated, if the price of the preceding trade is lower (higher) than the price of the current trade. If the price is the same for both trades, then the last recorded price change will be used to determine the direction. This method is very simple to implement and is the least demanding in terms of the information that must be available to the researcher since only the sequence of prices is needed for classification.

---

[6] See Lee and Ready (1991) and Hausman, Lo, and MacKinlay (1992).

- *Reverse tick test*: The current trade is classified as buyer (seller)-initiated, if the price of the next trade is lower (higher) than the price of the current trade. Despite its apparent similarity to the tick test, this method is seldom used in empirical applications, because it is known to be very inaccurate.
- *Midquote test*: The current trade is classified as buyer (seller)-initiated, if the current transaction price is higher (lower) than the last observed *midquote*, i.e. the arithmetic mean of the bid and the ask price. If the transaction price is equal to the midquote, the trade is left unclassified. This method obviously demands information on the prevailing bid and ask quotes at the time of the trade and thus cannot be used whenever the data set at hand contains only transactions.
- *Quote test*: The current trade is classified as buyer (seller) initiated, if the price of the current trade is closer to the last recorded ask (buy) quote. If the current price is at the midquote, then the last recorded price change is used to determine the direction. This method is a combination of the tick test and the midquote test and is applicable, whenever quote information is available.

Note that the quote test is a combination of the midquote test and the tick test, which is used only when trades occur at the midquote and therefore cannot be classified unambiguously by the midquote test. In many microstructure studies that employ data from the TAQ data base to estimate structural parameters of sequential trade models, including all of the studies we encountered in Chap. 3, a variation of the quote test, that accounts for NYSE specific features is used to determine the trade direction. The following 6-step algorithm is commonly referred to as the *Lee Ready*-method (LR-method):[7]

1. *Current quote match*: If the current quote has not been changed within the last 5 seconds before the transaction, trade direction is determined comparing the current quote to the trade price. Thus, when the price is equal to the ask (bid), the trade is classified as a buy (sell).
2. *Delayed quote match*: If the current quote is less than 5 seconds old, then the trade price is compared to the previous quote.
3. *Outside the spread match*: If trade prices are outside the spread as determined either in step one or two, the trade is classified on the basis of the distance of the price to the closest quote. Thus, when the price is greater than the ask (smaller than the bid), the trade is classified as a buy (sell).
4. *Midquote match*: If the price is at the midquote the tick test is used to classify the trade.
5. *Non-midquote match*: If the price is inside the spread, but not at the midquote the trade is classified according to its distance to the closest quote. Thus, when the price is closer to the ask (to the bid), the trade is classified as a buy (sell).

---

[7] It was originally proposed by Lee and Ready (1991). See also Lee and Radhakrishna (2000).

6. *Indeterminable*: If neither of the above matching conditions applies to a particular trade, it is classified as indeterminable. This is most frequently the case, when the first trade of the day occurs before the first quote is reported or the first trade occurs at the midquote before the first quote revision is reported, so the tick test cannot be applied.

The delayed quote match has been included in the algorithm for institutional reasons. At the NYSE trade and quote data comes from two different sources[8], so that time stamps of trades and quotes do not necessarily correspond to each other. Quotes are typically recorded approximately 5 seconds ahead of the trade prices on the NYSE.

### 5.2.2 Empirical Evidence on the Accuracy of Classification

The applicability and accuracy of classification algorithms in a variety of market settings has been studied by Lee and Ready (1991), Aitken and Frino (1996), Ellis, Michaely, and O'Hara (2000), Lee and Radhakrishna (2000), Odders-White (2000), Theissen (2000), and Finucane (2000). One should keep in mind, that the accuracy in most of these papers is not determined on the basis of an information set that would really allow one to classify all trades unambiguously. Rather some specific features of the data set at hand, which are not available in other high frequency data sets, are used to infer the 'true' trade direction. Typically, accuracy is assessed *assuming* that the additional information set makes the applied trade classification *exact*. The proportion of trades, that are classified by any algorithm in the same way as by the exact method is then used as a measure of accuracy. Of course, if the additional information used to compute the benchmark classification against which any other algorithm is compared is itself defective or incomplete, the results of this test procedure will be unreliable. Most of the studies focussing on the accuracy issue use data sets collected from the NYSE.

The evidence in Lee and Ready (1991) is based on a data set from the NYSE during 1988 for a sample of 150 firms. After correcting their quote change data set for inaccurate time stamps, they find that 92.1% of all buys at the ask and 90.0% of all sells at the bid are correctly classified by the simple tick test, assuming that the prevailing quote at the time of trading unambiguously identifies the true direction. While they find even higher proportions of correctly classified trades when prices are outside the spread, the performance of the tick test is difficult to evaluate, when trades occur inside the spread, since the prevailing quote cannot be used to identify the true direction in that case. However, they argue that most of the observed trades inside the spread result from the execution of *standing orders*, i.e. orders to buy or sell a certain

---

[8] Quote changes are usually recorded by the market maker's clerk, while employees of the NYSE record trade prices. Another source of discrepancies in the sequence in which quotes and trades appear is caused by different recording systems used to record trades and quotes. See Lee and Ready (1991) for details.

number of shares at the best available price over a certain time period. Assuming that order arrivals follow a simple Poisson process, they show that the tick test should classify these trades correctly in three of four possible cases.

The studies of Ellis, Michaely, and O'Hara (2000), Lee and Radhakrishna (2000) and Finucane (2000) are based on the *trades, orders, reports and quotes* (TORQ) data set, which contains information on transactions for a selection of 144 stocks traded on the NYSE. It covers a 3 month period beginning in November 1990. Beside the usual trade and quote information, it also contains information on the number and parties involved in each trade.[9] The TORQ database consists of 4 files, that contain different pieces of information. The consolidated trade (CT) and quote (CQ) files contain the same information on trades and quote revisions, that are available in the TAQ data set, and therefore need no further explanation.[10]

The usefulness of the TORQ data set for the assessment of classification accuracy stems from the inclusion of the remaining two files, the consolidated audit (CD) and the system order database (SOD) files. The CD is an expanded version of the CT file that additionally includes information on the order type (market, limit or other special order types) and the trader type (individual, institutional member or programmed trade). The SOD tape contains additional information on the origin of incoming orders. Specifically, it identifies all trades that are submitted over an electronic routing systems to the NYSE.[11] However, it does not contain any information on orders that are submitted to the trading floor non-electronically, such as trades submitted directly to floor brokers. Its use in determining the true direction is therefore somewhat limited.

Lee and Radhakrishna (2000) use only a fraction of 28.5% of the initial sample size of the TORQ data base for the assessment of classification accuracy. They drop all observations, for which they claim that the true direction cannot be determined unambiguously or may appear otherwise difficult to classify for the LR-method examined in their study. From the NYSE specific trade arrangements they identify three possible sources for misleading inferences with respect to the trade direction: Stopped orders, market crosses and orders that are either split up or batched at execution.[12]

*Stopped trades* occur, when the NYSE specialist stops an incoming market order and guarantees execution at the prevailing quote at a later date. The subsequent execution of a stopped order may appear either outside or inside

---

[9] See Hasbrouck (1992) for an exhaustive description of the TORQ database.

[10] See Sect. 5.1.

[11] There are three electronic order processing systems that provide access to the NYSE trading floor: The *SuperDot*, which is the main electronic order processing system for the NYSE, the *OARS* (Opening automated report system), which is used to submit orders for the opening batch auctions only, and the *ITS* (Intermarket trading system), which is used to submit orders between the NYSE and other stock exchanges in the U.S. See Sect. 2.2.3 for a description.

[12] See Sect. 2.2.2 for a complete description of NYSE order types.

the prevailing quote at execution time, thereby making the associated trade difficult to classify.[13] *Market crosses* are limit or market orders that are executed against another market or limit order, without direct participation of the market maker, so that again classification is difficult because both sides involved could have served as the initiator of the trade and market crosses are often executed inside the spread. *Order splits* make up for only 6% of the total number of all incoming market orders and thus the resulting bias should not be too severe. The percentage of batched orders however is three times as large (24%) in their data set and is increasing with order size, so that e.g. 56% of all trades involving more than 1900 shares have more than one participant on the active side.

Their final sample consists of 129,700 transactions, for which they find an overall success rate of 91.7% for the LR-method. The success rate is higher for trades with only a single participant on each side of the trade (93.3%) and lower for multiple participant trades, for which the true direction could be determined unambiguously (88.0%).

Finucane (2000) uses the TORQ data set to investigate the accuracy of the tick test, the reverse tick test and the LR-method. In contrast to Lee and Radhakrishna (2000) he retains many trades involving stopped orders and market crosses with at least one market order, but he still drops a large fraction of the initial sample either because one or both counter parties of the trade could not be identified using the information available in the audit file, or because the trade resulted from the submission of two limit orders. The initial sample size is reduced by roughly one half, leaving a total of 285,983 transactions in the final sample used to investigate the accuracy of classification.

He finds an overall success rate of 83.0% for the tick test, 72.1% for the reverse tick test and 84.4% for the LR-method, thus suggesting, that the LR-algorithm provides only a small improvement in accuracy over the simple tick test. However, these results confirm, that the reverse tick test is the least accurate method of classifying trades. Both, the tick test and the LR-method perform best when trades occur outside of the spread with success rates about 95%, while trades inside the spread but not on the midpoint are classified correctly about 83% of the times. Only roughly 66% of all trades that occur exactly at the midpoint are correctly classified by both methods. Trades that occur exactly at the ask or bid are being correctly classified in 87.6% (89.4%) of all cases by the tick test (LR-method). Further analysis shows that the most important factors that affect the accuracy of the LR-algorithm include the occurrence of market order crosses, stopped orders and trades on zero

---

[13] Note that stopped orders must not be confused with *stop orders*, which are limit orders that become market orders, once the transaction price reaches a pre-specified threshold value, see Sect. 2.2.2 for a discussion.

ticks[14], while the accuracy of the tick test is adversely affected when quote changes occur between trades. Taken together, his results suggest that all trades that are likely to receive price improvements over the posted bid and ask have a higher probability to be misclassified.

Using the same data set (TORQ), but a different procedure for determining the 'true' trade direction, Odders-White (2000) asses the accuracy of the tick test, the midquote test, and the LR-algorithm. She stresses the importance of a useful definition of the true trade direction, and discusses two possibilities that can be employed in practice. The *immediacy* definition emphasizes that the initiator of the trade is the party, who demands immediate execution of the trade. As a consequence of this definition traders who submit market orders or limit orders that immediately match an existing order in the opposite direction are labelled as initiators, while traders placing limit orders are typically viewed as non-initiators. This definition, implicitly used by Finucane (2000) and Lee and Radhakrishna (2000), does not help to classify many problematic transactions, e.g. when they result from direct matching of two market orders against each other. These may be classified however, when the *chronological* definition is used. According to this definition the investor who places his order last chronologically is the initiator.

Using the chronological definition she retains a large fraction of the initial TORQ data set for which an unambiguous determination of the true trade direction is possible. Only 25.1% of the initial sample size has to be dropped, leaving a total of 318,364 transaction used for subsequent assessment of accuracy. She finds an overall success rate of 78.6% for the tick test and 85.0% for the LR-method. The midquote test misclassifies only 9.1% of all trades, but it leaves another 16.0% of the trades unclassified because they occur at the midquote, so that the overall success rate for the midquote test is only 74.9%.

There is also evidence on the accuracy of classification methods for stock exchanges with a different market design than the NYSE. Using data for a two year period from the Australian Stock Exchange (ASX), which is a fully computerized trading system, Aitken and Frino (1996) find a smaller overall proportion of correctly classified trades. In their sample only 74.4% of all trades are given the right classification by the tick test, if essentially the same method as in Lee and Ready (1991) is used to identify the true trade direction.[15] However, excluding all zero ticks from their data set, raises the proportion of correct classifications to above 90 percent. In their data set trades occurring as zero ticks are the most commonly misclassified. Aitken and Frino (1996) report furthermore that seller initiated trades are more likely to be misspecified than buyer initiated trades, that the accuracy of the tick test

---

[14] *Zero ticks* are transactions that occur at the same price as the previous transaction.

[15] Aitken and Frino (1996) claim to be able to identify the true direction from their data set with a greater precision than Lee and Ready (1991), because their data set contains the exact chronological order of trades and quotes.

increases with trade size for buyer initiated trades and decreases in periods of high quote volatility or trending price behavior for both buyer and seller initiated trades.

Ellis, Michaely, and O'Hara (2000) compare the accuracy of the tick test, the midquote test and the quote test, using a data set that contains 313 stocks that had their initial public offering (IPO) between September 1996 and September 1997 and were consecutively traded on the NASDAQ trading platform.[16] Since they may identify the counter parties of each transaction, they are able to compare the performance of these classification algorithms to the trade direction as determined by the exact method. However, even so they have to exclude a significant proportion of their overall sample (24.6%), that are either trades between two market makers or between two brokers, so their true direction cannot be determined unambiguously. They find, that the quote test has the highest overall success rate of 81.1% of all remaining trades, followed by the tick test with 77.7% and the midquote test with 76.4%.

A closer look at the success rates for different types of trades reveals however, that the midquote test performs worse than the tick test, whenever trades occur either inside or outside the spread, but not at the bid or ask quote. The proximity of the transaction price to the prevailing quotes turns out to be the single most important determinant affecting trade misclassifications in their sample. Therefore they propose to use the midquote test only for trades that occur exactly at the ask or at the bid, and to classify all remaining trades employing the tick test. For this alternative algorithm they find a overall success rate of 81.9%, which appears to be only a slight improvement in accuracy compared to the quote test.

Theissen (2000) uses a sample of 15 selected shares traded on the Frankfurt Stock Exchange (FSE) during one month in 1996 to compare the tick test and the quote test. His data set allows him to determine the true trade direction for almost 90% of the initial sample size, by classifying all trades to be buys (sells) if the market maker on the FSE sold (bought) shares. His findings show that the tick test with a success rate of 72.2% is almost as accurate as the classification provided by the quote test with an overall success rate of 72.8%.[17] The accuracy of classification tends to be higher for more liquid stocks, and again trades occurring on a zero tick appear to be more problematic than trades occurring on a non-zero tick.

---

[16] Their initial sample contains 2,433,019 trades and 627,370 quotes.

[17] Evaluation of the quote test is based on 9,449 observations, while the evaluation of the tick test is based on 9,124 transactions only. This discrepancy arises, because the classification of the trades is based on all observations of the initial sample, including those for which the true direction could not be determined. He drops all transactions for which either the true direction is unknown or which are classified as indeterminable in a second step.

### 5.2.3 Classification of Trades

The classification of the trades in our sample was conducted using the LR-method, i.e. all transaction prices were compared to the bid and ask quotes lagged by 5 seconds, thus applying the delayed quote match as suggested by Lee and Ready (1991). If the price of a particular transaction was equal to the midquote, the tick test was applied. Given the fact that we have no exact information on whether any particular transaction was buyer or seller initiated, there is practically no way to completely avoid misclassification as there is no alternative to applying one of the algorithms discussed in the preceding Sections. The LR-method is tailor-made for the use of transaction data collected on the NYSE and appears to be the most accurate of the proposed algorithms and so we decided to use this method. Virtually all of the empirical studies we discussed in Chap. 3 that estimate variants of the sequential trade model based on transaction data from the NYSE employ the same algorithm for trade classification as well, so the comparability of our approach to these studies is another benefit of using the LR-method.

The discussion in the preceding Section also indicated that the accuracy of classification is most intimately related to the proximity of transaction prices to the prevailing bid and ask quotes, with trades occurring outside the spread or at the bid or ask typically being more accurately classified by the LR-method than trades inside the spread. Furthermore, the classification of zero-ticks appears to be more problematic than trades occurring on non zero-ticks. In order to get some indication of the potential severity of misclassification in our sample we present some summary statistics with regard to the classification step. Table 5.2 contains the relative frequencies of observations in our sample that belong to a group defined by the relevant characteristics.

With the exception of IBM, all of the stocks in our sample exhibit a slightly higher proportion of buys (53%) than sells (47%). Only 0.25% of all transactions were left unclassified. Such indeterminable transactions typically occur at the beginning of the trading day. If trades occur before the first quotes are posted or the transaction price is exactly at the midpoint of the bid and ask quotes, the LR-method compares the actual trade price to lagged trade prices, but if the trading day starts with a sequence of trades at the same price, it is not possible to classify them unambiguously. Note that at the NYSE each trading day starts with a batch auction conducted by the delegated market maker, so an unknown number of trades after the open may result from these batch auctions, rather than from continuous trading.

The fraction of trades that are potentially hard to classify appears to be relatively fair sized in our sample. About 79% of all trades are either exactly at the bid or the ask, thus not appearing to be problematic with respect to the classification issue. Given that the most sensitive trades are those that receive some kind of price improvement and are therefore recorded inside the prevailing bid-ask spread, we find only a fraction of about 21% of all trades in our sample belong to this group, which predominantly consists of trades that occurred exactly at the midpoint (98%). Around 75% of all observations in our sample occurred on a zero tick, but the bulk of these trades occurred either at the ask (35%) or at the bid (40%), so the classification of these zero ticks should not be problematic as well. However, one quarter of the zero tick trades were exactly at the midpoint, so the tick test had to be applied to classify them. Noting that the midpoint trades were at the same time most likely zero ticks (92%), we find that this group consisting of roughly 19% of all trades is potentially at the highest risk of being misclassified.

**Table 5.2.** Classification of trades

| Type of trade | Price location | | | | | | | P |
|---|---|---|---|---|---|---|---|---|
| | $p < b$ | $p = b$ | $b < p < m$ | $p = m$ | $m < p < a$ | $p = a$ | $a < p$ | |
| **Boeing** | | | | | | | | |
| Non-zero tick | 0.03 | 13.85 | 0.04 | 2.80 | 0.00 | 12.95 | 0.29 | 29.95 |
| Zero tick | 0.01 | 18.44 | 0.11 | 28.75 | 0.12 | 22.56 | 0.07 | 70.05 |
| Indeterminable | 0.00 | 0.00 | 0.00 | 0.13 | 0.00 | 0.00 | 0.30 | 0.44 |
| Sell | 0.04 | 32.28 | 0.15 | 12.17 | 0.00 | 0.00 | 0.03 | 44.67 |
| Buy | 0.00 | 0.00 | 0.00 | 19.24 | 0.12 | 35.51 | 0.03 | 54.90 |
| P | 0.04 | 32.28 | 0.15 | 31.54 | 0.12 | 35.51 | 0.36 | 100 |
| **Disney** | | | | | | | | |
| Non-zero tick | 0.06 | 12.53 | 0.06 | 1.20 | 0.05 | 12.75 | 0.08 | 26.74 |
| Zero tick | 0.00 | 18.99 | 0.08 | 18.34 | 0.31 | 35.50 | 0.04 | 73.26 |
| Indeterminable | 0.00 | 0.00 | 0.00 | 0.04 | 0.00 | 0.00 | 0.10 | 0.14 |
| Sell | 0.06 | 31.51 | 0.14 | 11.06 | 0.00 | 0.00 | 0.00 | 42.78 |
| Buy | 0.00 | 0.00 | 0.00 | 8.44 | 0.36 | 48.26 | 0.02 | 57.08 |
| P | 0.06 | 31.51 | 0.14 | 19.54 | 0.36 | 48.26 | 0.12 | 100 |
| **IBM** | | | | | | | | |
| Non-zero tick | 0.13 | 10.52 | 0.16 | 2.64 | 0.14 | 10.48 | 0.11 | 24.17 |
| Zero tick | 0.01 | 30.97 | 0.17 | 21.17 | 0.43 | 23.01 | 0.08 | 75.83 |
| Indeterminable | 0.00 | 0.00 | 0.00 | 0.05 | 0.00 | 0.00 | 0.16 | 0.21 |
| Sell | 0.14 | 41.49 | 0.32 | 10.81 | 0.00 | 0.00 | 0.00 | 52.75 |
| Buy | 0.00 | 0.00 | 0.00 | 12.94 | 0.56 | 33.49 | 0.03 | 47.03 |
| P | 0.14 | 41.49 | 0.32 | 23.80 | 0.56 | 33.49 | 0.19 | 100 |
| **Coca-Cola** | | | | | | | | |
| Non-zero tick | 0.13 | 15.11 | 0.05 | 0.21 | 0.05 | 15.07 | 0.15 | 30.77 |
| Zero tick | 0.01 | 28.99 | 0.00 | 7.91 | 0.00 | 32.27 | 0.04 | 69.23 |
| Indeterminable | 0.00 | 0.00 | 0.00 | 0.03 | 0.00 | 0.00 | 0.13 | 0.17 |
| Sell | 0.14 | 44.11 | 0.05 | 1.89 | 0.00 | 0.00 | 0.00 | 46.18 |
| Buy | 0.00 | 0.00 | 0.00 | 6.20 | 0.05 | 47.34 | 0.06 | 53.65 |
| P | 0.14 | 44.11 | 0.05 | 8.12 | 0.05 | 47.34 | 0.19 | 100 |
| **Exxon** | | | | | | | | |
| Non-zero tick | 0.01 | 8.37 | 0.01 | 0.68 | 0.01 | 8.34 | 0.13 | 17.55 |
| Zero tick | 0.00 | 27.03 | 0.01 | 20.30 | 0.00 | 35.01 | 0.11 | 82.45 |
| Indeterminable | 0.00 | 0.00 | 0.00 | 0.18 | 0.00 | 0.00 | 0.21 | 0.39 |
| Sell | 0.01 | 35.40 | 0.02 | 9.30 | 0.00 | 0.00 | 0.00 | 44.72 |
| Buy | 0.00 | 0.00 | 0.00 | 11.50 | 0.01 | 43.35 | 0.02 | 54.89 |
| P | 0.01 | 35.40 | 0.02 | 20.98 | 0.01 | 43.35 | 0.23 | 100 |
| **All stocks** | | | | | | | | |
| Non-zero tick | 0.08 | 11.98 | 0.07 | 1.52 | 0.06 | 11.87 | 0.14 | 25.74 |
| Zero tick | 0.01 | 25.95 | 0.08 | 18.64 | 0.19 | 29.32 | 0.07 | 74.26 |
| Indeterminable | 0.00 | 0.00 | 0.00 | 0.08 | 0.00 | 0.00 | 0.17 | 0.25 |
| Sell | 0.09 | 37.93 | 0.15 | 8.80 | 0.00 | 0.00 | 0.00 | 46.98 |
| Buy | 0.00 | 0.00 | 0.00 | 11.28 | 0.25 | 41.20 | 0.03 | 52.76 |
| P | 0.09 | 37.93 | 0.15 | 20.16 | 0.25 | 41.20 | 0.21 | 100 |

$p$ = transaction price, $b$ = bid, $m$ = midquote, $a$ = ask. The entries in Table 5.2 are the percentages of trades, falling into the specified classes.

## 5.3 Descriptive Statistics

The final step of our data preparation procedure consists of counting the number of buy and sell transactions over some predefined time interval. We employ a basic time interval of 5 minutes which is often used in intraday studies of high frequency data sets[18] and has also been used in previous empirical studies on sequential trade models to define no-trade events.[19] Thus, we set $\Delta t$ equal to 1 in the following. Table 5.3 contains summary statistics for our count data set consisting of the number of buys respectively sells per five minute interval.

**Table 5.3.** Descriptive statistics

| Statistic | BA | DIS | IBM | KO | XON |
|---|---|---|---|---|---|
| | | Buys | | | |
| $T$ | 1716 | 1716 | 1716 | 1716 | 1716 |
| Mean | 2.426 | 3.213 | 4.233 | 3.747 | 3.028 |
| Variance | 7.434 | 9.149 | 23.361 | 11.498 | 9.443 |
| Skewness | 1.806 | 1.814 | 1.681 | 1.804 | 1.338 |
| Kurtosis | 4.406 | 6.402 | 3.246 | 5.122 | 1.959 |
| Minimum | 0 | 0 | 0 | 0 | 0 |
| Median | 2 | 3 | 3 | 3 | 2 |
| Maximum | 19 | 28 | 29 | 25 | 18 |
| Mode | 0 | 0 | 0 | 2 | 0 |
| $\rho$ | 0.023 | -0.023 | -0.058 | -0.025 | -0.008 |
| $\chi^2_{LB}$ | 542.4 | 305.7 | 1611.3 | 205.8 | 515.5 |
| | | Sells | | | |
| $T$ | 1716 | 1716 | 1716 | 1716 | 1716 |
| Mean | 1.974 | 2.408 | 4.748 | 3.226 | 2.467 |
| Variance | 4.564 | 5.728 | 18.638 | 7.654 | 6.306 |
| Skewness | 1.478 | 1.653 | 1.431 | 1.401 | 1.553 |
| Kurtosis | 2.461 | 4.232 | 2.595 | 2.642 | 3.377 |
| Minimum | 0 | 0 | 0 | 0 | 0 |
| Median | 1 | 2 | 4 | 3 | 2 |
| Maximum | 13 | 18 | 29 | 18 | 17 |
| Mode | 0 | 0 | 0 | 2 | 0 |
| $\rho$ | 0.042 | -0.027 | -0.008 | -0.015 | 0.018 |
| $\chi^2_{LB}$ | 324.4 | 383.9 | 1227.2 | 385.6 | 269.0 |

$T$ is the number of observations, $\rho$ is the estimated autocorrelation coefficient for 50 lags. $\chi^2_{LB}$ is the Ljung-Box statistic for 50 lags. The corresponding tabulated 99% (95%) critical value for 50 degrees of freedom is 76.2 (67.5).

All time series are highly overdispersed, with variance exceeding the mean by factors between 2.31 (BA sells) and 5.52 (KO buys). Furthermore, all ten series have higher skewness coefficients, indicating that the distribution is extremely right skewed, and have fatter tails than an ordinary Poisson distri-

---

[18] See e.g. Andersen and Bollerslev (1997).
[19] See e.g. Easley, Kiefer, and O'Hara (1997b).

bution with the same mean.[20] In addition, the values of the Ljung-Box test statistic exceed their corresponding critical values by far. This confirms our earlier finding, that time series of transaction data exhibit significant auto-correlation for long lag lengths. The sinusoidal patterns of the autocorrelation function (ACF) for our sample data shown in Fig. 5.1 reveal furthermore, that the time series appear to contain some periodic components as well. Note, that the sample ACF appears to have peaks at each integer multiple of 78, which is exactly equal to the number of 5 minute intervals per trading day. However, there are differences in the regularity of these patterns between the stocks in our sample, with the IBM stock having the most pronounced and regular sinusoidal pattern. These patterns appear to be related to intradaily seasonal movements in the trading activity, with more or less pronounced short run components imposed on the seasonal pattern that distort the picture.

Figure 5.2 contains estimates of the time of day function for the time series in our sample. These estimates were obtained by computing the mean number of trading events observed over successive 5 minute intervals. It is a well known stylized fact that trading activity at the NYSE evolves during the regular trading day (i.e. when the initial batch auction phase is neglected) according to a U-shaped pattern, i.e. it shows peaks at the opening and the close. When the auction period is included, the trading process features an asymmetry in the sense, that trading activity is increasing and takes on the U-shape after completion of the batch auction.

During the opening auction, the designated market maker sets a price in order to maximize the transaction volume. After the price is fixed and orders are executed, transactions are being recorded. When the auction is completed, continuous trading begins. Because of a delay of the open that may occur from time to time, the first regular daily transaction may be recorded some minutes after the official open at 9:30 a.m. Hence, we expect an initially low trading activity at the open that increases quickly before the idiosyncratic U-shape pattern is assumed. The typical plot of the time of day function in Fig. 5.2 is indeed roughly U-shaped, with a dip in the first interval at the beginning of trading day. This dip in the first interval appears to be related to the batch auction conducted at the beginning of the trading day.

In addition we present plots of the sample densities of the trading events in Fig. 5.3. The expected density included for comparison in Fig. 5.3 is the Poisson evaluated using the sample means of the event counts as estimates for the parameter $\lambda$. The plot reveals that the distribution of the number of buys and sells per five minute interval appears to be sharply distinct from the shape that a one component Poisson density with the same mean would exhibit. With the exception of the KO buys and sells, all the sample densities have their mode at zero and the frequency of a higher number of events is declining steadily as we move away from the mode, which is typical for right

---

[20] The Poisson distribution has skewness coefficient $\alpha_3 = \frac{\mu_3}{(\sqrt{\mu_2})^3}$ ($\mu_s$ is the $s$-th central moment) equal to $\lambda^{-0.5}$ and excess kurtosis $\alpha_4 = \frac{\mu_4}{(\mu_2)^2} - 3$ equal to $\lambda^{-1}$.

skewed distributions. None of the densities appears to have a pronounced bimodal shape. Furthermore, the fat tails of the sampling densities are easily recognizable by comparing the relative frequency of trading events with values of 10 and greater. The densities for the KO trade events both have a mode at 2, but are otherwise comparable to the densities of the other stocks.

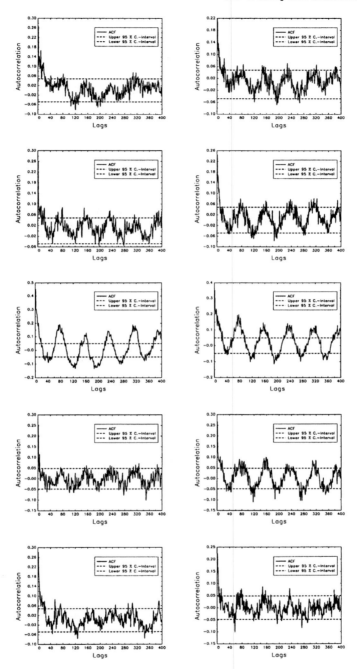

**Fig. 5.1.** Autocorrelation function. Left panel: Buys, right panel: Sells. From top to bottom: Boeing, Disney, IBM, Coca-Cola, Exxon.

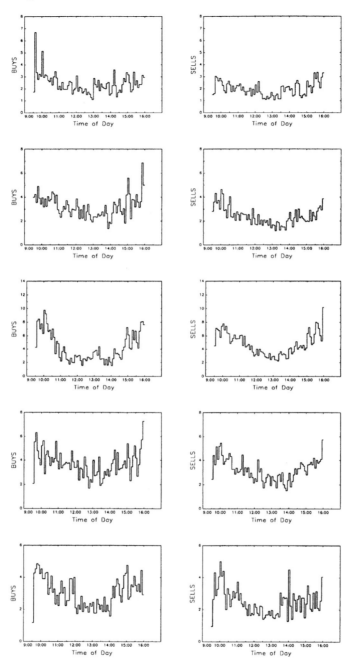

**Fig. 5.2.** Time of day function. Left panel: Buys, right panel: Sells. From top to bottom: Boeing, Disney, IBM, Coca-Cola, Exxon.

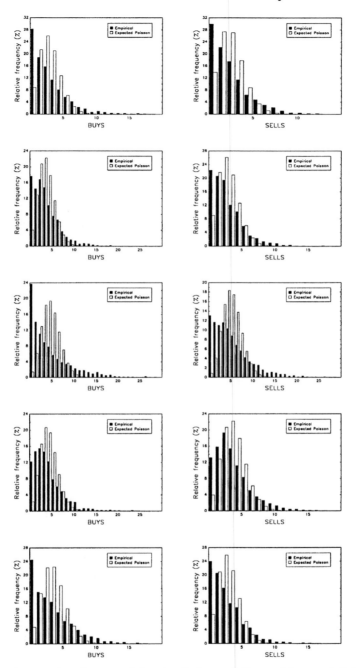

**Fig. 5.3.** Empirical and expected density. Left panel: Buys, right panel: Sells. From top to bottom: Boeing, Disney, IBM, Coca-Cola, Exxon.

## 5.4 Estimation Results

### 5.4.1 Model Selection

In a first step towards finding an appropriate statistical model for our time series, we estimated a number of unrestricted bivariate mixture and Markov switching regression models. Given the huge number of possible model specifications, we had to restrict our attention to a subset of plausible models, both in terms of the empirical features of the time series in our sample and in terms of the underlying economic model. We therefore estimate 2, 3, and 4 regime static mixture, denoted in the following by the token M, and Markov switching (MS) regression models based on the Poisson (P) and on the negative binomial (N) density. This makes a total of four different model specifications in terms of the marginal density and the specification of the regime variable for each given number of states. For each level of regime components we furthermore combined these four different models with another four different specifications for the regime dependent regression function. The general specification of the regime dependent regression function was

$$\ln \lambda_{tj} = \omega^{(j)} + \sum_{s=1}^{S} \left[ \zeta_s^{(j)} \cdot \cos\left(s \cdot h(\tau_t)\right) + \eta_s^{(j)} \cdot \sin\left(s \cdot h(\tau_t)\right) \right] \qquad (5.1)$$

$$+\delta^{(j)} \cdot I(\tau_{\min}) + \phi_B^{(j)} \cdot \ln B_{t-1}^* + \phi_S^{(j)} \cdot \ln S_{t-1}^*,$$

where the normalizing function $h(\tau)$ is given by

$$h(\tau) = 2\pi \cdot \frac{\tau - \tau_{\min}}{\tau_{\max} - \tau_{\min}},$$

$\tau_{\min}$ ($\tau_{\max}$) is the time of day at which trading begins (ends) at the NYSE, $I(\tau_{\min})$ is an indicator dummy variable for the first trading interval (9:30 - 9:35), and the lagged dependent variables $B_t$ and $S_t$ are defined by $y_{t-m}^* = \max(y_{t-m}, 0.5)$ in order to make the logarithmic transformation of zero observations possible. The four different specifications of the regression function are obtained by imposing zero restrictions on some of the parameters in (5.1). Our first specification, denoted as model C, will include only a constant term, while the second specification S, will additionally include sine and cosine terms as well as the dummy variable $I(\tau_{\min})$ for the first trading interval as a means to control for the seasonal component of the time series.

In all specifications, in which a seasonal component is included in the regression function, the truncation lag of the Fourier series $s$ is set equal to 1. Though the use of this specification for the seasonal component is motivated by the SNP estimator introduced in Sect. 4.1.5 that would typically include higher order polynomials of the function $h(\tau)$ as additional regressors, we found by way of experimentation that for our log-linear regression function the inclusion of polynomial terms almost inevitably led to explosive regime specific forecasts of the event counts. Given this unstable behavior, we restricted ourselves to the parsimonious specification of (5.1), with the indicator dummy variable $I(t_{\min})$ included as an additional regressor variable, motivated by the fact that the trading process at the NYSE starts with a batch auction that overlaps into the first 5 minute interval from time to time before it switches to continuous trading, so the observed number of trades in our data set is typically lower during the first interval than in consecutive intervals.

The last two specifications for the regime specific regression function include lags of the dependent variables as additional regressors. In the model specification termed SAR, we included only lags of the dependent variable, so that the conditional mean for the time series of buys will only include lags of $B_t$ et vice versa. The last model specification SVAR will include lags of the dependent variable of the other equation as well, so that the conditional mean for the buys will depend on lags of $B_t$ as well as on lags of $S_t$ as in an ordinary vector autoregressive model.

Thus, we end up with 16 different specifications for each level of regime components, making a total of 48 different models that we estimated for each of the five stocks in our sample. Estimation of the specified model specifications was conducted employing the EM-algorithm as laid out in Sect. 4.1.1. The M-step was conducted by separate maximization of the marginal likelihood contributions for each regime and equation based on weighted Poisson respectively Negative binomial regressions with weights given by the time variable regime probabilities obtained in the E-step, i.e. $\hat{\xi}_t^{(j)}$ when a static mixture model and $\hat{\xi}_{t|T}^{(j)}$ when a Markov switching model was specified.[21] The iterations defining the EM-algorithm were repeated until the maximal absolute change of the parameter vector $\theta$ was smaller than $\epsilon = 10^{-6}$ or until 500 iterations were performed, whatever came first. Standard errors for the parameter estimates were calculated based on the analytical gradients and Hessian matrix of the expected complete log-likelihood function.

As discussed in Sect. 4.1.1, the EM-algorithm is capable of finding local maxima of the incomplete log-likelihood function. It is also well known, that the likelihood function of a mixture model always has several local maxima, some of which may be found by simply relabelling the regimes after convergence. In order to ensure, that a global maximum has indeed been found, we therefore estimated each model specification a number of times with different start values. Apart from a few exceptions, this procedure always led to the same parameter estimates (except for a permutation of the regime labels) from different start values. Usually, we would repeat each estimation 25 times

---

[21] Maximization of the marginal likelihood contributions was performed based on analytical expressions for the gradient and Hessian matrix employing the Davidon-Fletcher-Powell algorithm, which is available in the econometric software package GAUSS through the Constrained Maximum Likelihood (CML)-library. All other algorithms used in this study were programmed by the author.

with different start values in order to ensure that the algorithm converged
to the same maxima in each step, but when different maxima were found we
repeat it up to 100 times using random start values[22] and take the parameter
estimates associated with the maximum value of the incomplete log-likelihood
function found in these repetitions as the ML-estimate.

Our primary goal is to find the most parsimonious model specification that
yields a satisfactory description of the sample data. Thus, our choice is based
on the information criteria $AIC$ and $BIC$. In addition we conducted for each
estimated model specification tests for residual autocorrelation, based on the
Ljung-Box test for the Pearson residuals and for goodness of fit, based on
the conditional moment test. Both tests were described in Sect. 4.2 in greater
detail. Tables 5.4 through 5.8 contain the results of this step, specifically the
values of the incomplete log-likelihood function, the information criteria, and
the p-values of the test statistics.[23] It turns out that the $BIC$ selects the
MS-N-SAR 3-regime specification in all cases, while the $AIC$ selects in all
cases either the MS-N-SVAR 3-regime (KO), or a 4 regime model, based on
a MS-N-SAR (BA, XON) or MS-N-SVAR (IBM, DIS) specification for the
conditional mean, but never a more parsimonious model than the $BIC$. This
behavior might have been expected however, since the $AIC$ is known to have
a bias towards overestimating the order of a model.[24]

---

[22] By way of trial and error, we found that the critical quantities in such cases
appeared to be the start values of the mixing probabilities $\pi_j$ and transition
probabilities $p_{ji}$, so we draw an appropriate number of pseudo random variables
from an uniform distribution on the interval $[0, 1]$, standardized them so that
the $\pi_j$ or the elements in each row of the transition matrix $P$ sum up to one as
required, and match them with start values for the regression coefficients. These
were obtained from weighted OLS-regressions of the logarithm of the dependent
variable on the associated regressor variables conducted separately for each regime
and dependent variable. Start values for the dispersion parameter $\alpha$ of the negative binomial distribution have been computed employing a moment estimator
based on the residuals from these regressions, see Cameron and Trivedi (1998),
p. 65. Zero observations of the dependent variables were replaced by the value
0.5 in order to make the logarithmic transformation possible. The weights used
in these auxiliary regressions were pseudo random variables as well, drawn from
an uniform distribution on the unit interval and standardized so that their sum
was equal to one for each observation and their mean was equal to the mixing or
stationary probabilities $\pi_j$.

[23] The p-value of a test statistic is equal to the maximal significance level, at which
the maintained Null hypothesis may be rejected. If e.g. $p(\chi^2_{LB}) = 0.04$, then the
corresponding Null may be rejected at the 5% significance level, but not at the
1% level.

[24] See Sawa (1978), Wang, Puterman, Cockburn, and Le (1996), and Wang, Cockburn, and Puterman (1998).

**Table 5.4.** Model selection BA

| Model | $\ln \mathcal{L}_I(\theta)$ | AIC | BIC | $p(\chi^2_{LB})$ | | $p(\chi^2_{CM})$ | |
|---|---|---|---|---|---|---|---|
| | | | | Buys | Sells | Buys | Sells |

|  |
|---|
| **2 Regime specification** |

| Model | $\ln \mathcal{L}_I(\theta)$ | AIC | BIC | Buys | Sells | Buys | Sells |
|---|---|---|---|---|---|---|---|
| M-P-C | -7151.1 | 14314.2 | 14346.9 | 0.0000 | 0.0000 | 0.0000 | 0.0000 |
| M-P-S | -7027.5 | 14091.0 | 14189.1 | 0.0000 | 0.0000 | 0.0000 | 0.0000 |
| M-P-SAR | -6850.1 | 13744.3 | 13864.1 | 0.0000 | 0.0000 | 0.0000 | 0.0000 |
| M-P-SVAR | -6840.6 | 13733.2 | 13874.8 | 0.0000 | 0.0001 | 0.0000 | 0.0000 |
| M-N-C | -6772.1 | 13564.2 | 13618.7 | 0.0000 | 0.0000 | 0.0001 | 0.0000 |
| M-N-S | -6713.0 | 13469.9 | 13589.8 | 0.0000 | 0.0000 | 0.0013 | 0.0004 |
| M-N-SAR | -6598.8 | 13249.7 | 13391.3 | 0.0000 | 0.0000 | 0.0043 | 0.2181 |
| M-N-SVAR | -6591.2 | 13242.4 | 13405.8 | 0.0000 | 0.0001 | 0.0085 | 0.1563 |
| MS-P-C | -7121.2 | 14258.4 | 14302.0 | 0.0000 | 0.0000 | 0.0000 | 0.0000 |
| MS-P-S | -6998.3 | 14036.7 | 14145.6 | 0.0000 | 0.0000 | 0.0000 | 0.0000 |
| MS-P-SAR | -6850.0 | 13748.1 | 13878.8 | 0.0000 | 0.0000 | 0.0000 | 0.0000 |
| MS-P-SVAR | -6836.5 | 13729.0 | 13881.6 | 0.0000 | 0.0000 | 0.0000 | 0.0000 |
| MS-N-C | -6750.7 | 13525.4 | 13590.7 | 0.0000 | 0.0000 | 0.0001 | 0.0006 |
| MS-N-S | -6704.1 | 13456.2 | 13586.9 | 0.0000 | 0.0000 | 0.0110 | 0.0227 |
| MS-N-SAR | -6598.3 | 13252.6 | 13405.1 | 0.0000 | 0.0000 | 0.0053 | 0.2418 |
| MS-N-SVAR | -6591.4 | 13246.7 | 13421.1 | 0.0000 | 0.0001 | 0.0069 | 0.0196 |

|  |
|---|
| **3 Regime specification** |

| Model | $\ln \mathcal{L}_I(\theta)$ | AIC | BIC | Buys | Sells | Buys | Sells |
|---|---|---|---|---|---|---|---|
| M-P-C | -6914.4 | 13846.7 | 13895.7 | 0.0000 | 0.0000 | 0.0000 | 0.0000 |
| M-P-S | -6833.1 | 13720.1 | 13867.2 | 0.0000 | 0.0000 | 0.0000 | 0.0000 |
| M-P-SAR | -6691.4 | 13448.9 | 13628.7 | 0.0000 | 0.0000 | 0.0000 | 0.0000 |
| M-P-SVAR | -6663.9 | 13405.9 | 13618.3 | 0.0000 | 0.0001 | 0.0000 | 0.0000 |
| M-N-C | -6772.0 | 13574.0 | 13655.7 | 0.0000 | 0.0000 | 0.0000 | 0.0000 |
| M-N-S | -6694.0 | 13454.1 | 13633.8 | 0.0000 | 0.0000 | 0.0005 | 0.2928 |
| M-N-SAR | -6572.7 | 13223.4 | 13435.8 | 0.0000 | 0.0000 | 0.0579 | 0.3640 |
| M-N-SVAR | -6563.5 | 13217.0 | 13462.1 | 0.0000 | 0.0001 | 0.0032 | 0.1701 |
| MS-P-C | -6808.1 | 13646.1 | 13727.8 | 0.0000 | 0.0005 | 0.0000 | 0.0000 |
| MS-P-S | -6750.3 | 13566.5 | 13746.3 | 0.0000 | 0.0000 | 0.0000 | 0.0000 |
| MS-P-SAR | -6679.1 | 13436.2 | 13648.6 | 0.0000 | 0.0000 | 0.0000 | 0.0000 |
| MS-P-SVAR | -6652.4 | 13394.8 | 13639.9 | 0.0000 | 0.0001 | 0.0000 | 0.0000 |
| MS-N-C | -6630.2 | 13302.3 | 13416.7 | 0.0011 | 0.1925 | 0.0000 | 0.0077 |
| MS-N-S | -6595.1 | 13268.1 | 13480.6 | 0.0000 | 0.1364 | 0.0001 | 0.0202 |
| MS-N-SAR | -6545.9 | 13181.7 | 13426.9 | 0.0877 | 0.3328 | 0.0002 | 0.1426 |
| MS-N-SVAR | -6557.2 | 13216.3 | 13494.1 | 0.1925 | 0.2063 | 0.0020 | 0.0568 |

|  |
|---|
| **4 Regime specification** |

| Model | $\ln \mathcal{L}_I(\theta)$ | AIC | BIC | Buys | Sells | Buys | Sells |
|---|---|---|---|---|---|---|---|
| M-P-C | -6820.0 | 13664.1 | 13729.4 | 0.0000 | 0.0000 | 0.0000 | 0.0000 |
| M-P-S | -6779.3 | 13630.6 | 13826.8 | 0.0000 | 0.0000 | 0.0000 | 0.0000 |
| M-P-SAR | -6607.1 | 13302.2 | 13541.9 | 0.0000 | 0.0000 | 0.0133 | 0.0000 |
| M-P-SVAR | -6599.3 | 13302.6 | 13585.8 | 0.0000 | 0.0001 | 0.0000 | 0.0002 |
| M-N-C | -6789.4 | 13618.8 | 13727.8 | 0.0000 | 0.0000 | 0.0000 | 0.0000 |
| M-N-S | -6692.5 | 13473.0 | 13712.7 | 0.0000 | 0.0000 | 0.0000 | 0.0401 |
| M-N-SAR | -6572.4 | 13248.7 | 13532.0 | 0.0000 | 0.0000 | 0.0000 | 0.0034 |
| M-N-SVAR | -6552.7 | 13225.5 | 13552.3 | 0.0000 | 0.0001 | 0.0061 | 0.0107 |
| MS-P-C | -6660.1 | 13368.2 | 13499.0 | 0.0024 | 0.1229 | 0.0000 | 0.0000 |
| MS-P-S | -6659.5 | 13414.9 | 13676.4 | 0.0001 | 0.0095 | 0.0000 | 0.0000 |
| MS-P-SAR | -6560.2 | 13232.4 | 13537.4 | 0.0000 | 0.0166 | 0.0000 | 0.0000 |
| MS-P-SVAR | -6561.6 | 13251.2 | 13599.8 | 0.0000 | 0.0005 | 0.0000 | 0.0000 |
| MS-N-C | -6598.5 | 13261.0 | 13435.3 | 0.0188 | 0.4094 | 0.0001 | 0.5570 |
| MS-N-S | -6540.6 | 13193.2 | 13498.3 | 0.0127 | 0.3226 | 0.0001 | 0.0407 |
| MS-N-SAR | -6521.9 | 13171.8 | 13520.4 | 0.4892 | 0.4926 | 0.0000 | 0.0483 |
| MS-N-SVAR | -6541.2 | 13226.4 | 13618.6 | 0.4505 | 0.2913 | 0.0006 | 0.0303 |

$\chi^2_{LB}$ is computed for 50 lags. $\chi^2_{CM}$ is computed for 12 cells, $(0,...,10, 11+)$.
$p(\chi^2)$ denotes the p-values of the corresponding test statistics. M = Static mixture model, MS = Markov switching model, P = Poisson model, N = Negbin 2 model, C = Constants only model, S = Seasonal model, SAR = Seasonal autoregressive model, SVAR = Seasonal vector autoregressive model.

**Table 5.5.** Model selection DIS

| Model | $\ln \mathcal{L}_I(\theta)$ | AIC | BIC | $p(\chi^2_{LB})$ | | $p(\chi^2_{CM})$ | |
|---|---|---|---|---|---|---|---|
| | | | | Buys | Sells | Buys | Sells |

**2 Regime specification**

| Model | $\ln \mathcal{L}_I(\theta)$ | AIC | BIC | Buys | Sells | Buys | Sells |
|---|---|---|---|---|---|---|---|
| M-P-C | -7739.8 | 15491.7 | 15524.3 | 0.0000 | 0.0000 | 0.0000 | 0.0000 |
| M-P-S | -7538.6 | 15113.1 | 15211.2 | 0.0000 | 0.0000 | 0.0000 | 0.0000 |
| M-P-SAR | -7409.2 | 14862.3 | 14982.2 | 0.0002 | 0.0057 | 0.0000 | 0.0000 |
| M-P-SVAR | -7402.4 | 14856.8 | 14998.4 | 0.0034 | 0.0077 | 0.0000 | 0.0000 |
| M-N-C | -7380.7 | 14781.4 | 14835.9 | 0.0000 | 0.0000 | 0.0000 | 0.0000 |
| M-N-S | -7283.6 | 14611.2 | 14731.1 | 0.0000 | 0.0000 | 0.0000 | 0.0372 |
| M-N-SAR | -7200.3 | 14452.7 | 14594.3 | 0.0007 | 0.0051 | 0.0000 | 0.0719 |
| M-N-SVAR | -7194.7 | 14449.3 | 14612.7 | 0.0215 | 0.0039 | 0.0001 | 0.1197 |
| MS-P-C | -7708.0 | 15432.0 | 15475.6 | 0.0000 | 0.0000 | 0.0000 | 0.0000 |
| MS-P-S | -7505.7 | 15051.4 | 15160.4 | 0.0000 | 0.0000 | 0.0000 | 0.0000 |
| MS-P-SAR | -7408.0 | 14864.1 | 14994.8 | 0.0002 | 0.0034 | 0.0000 | 0.0000 |
| MS-P-SVAR | -7395.4 | 14846.8 | 14999.3 | 0.0055 | 0.0066 | 0.0000 | 0.0000 |
| MS-N-C | -7355.6 | 14735.2 | 14800.5 | 0.0000 | 0.0000 | 0.0000 | 0.0005 |
| MS-N-S | -7253.3 | 14554.6 | 14685.3 | 0.0000 | 0.0000 | 0.0000 | 0.0741 |
| MS-N-SAR | -7199.8 | 14455.6 | 14608.1 | 0.0005 | 0.0045 | 0.0000 | 0.0370 |
| MS-N-SVAR | -7190.7 | 14445.5 | 14619.8 | 0.0125 | 0.0054 | 0.0003 | 0.1096 |

**3 Regime specification**

| Model | $\ln \mathcal{L}_I(\theta)$ | AIC | BIC | Buys | Sells | Buys | Sells |
|---|---|---|---|---|---|---|---|
| M-P-C | -7526.1 | 15070.3 | 15119.3 | 0.0000 | 0.0000 | 0.0000 | 0.0000 |
| M-P-S | -7387.6 | 14829.1 | 14976.2 | 0.0000 | 0.0000 | 0.0000 | 0.0000 |
| M-P-SAR | -7268.9 | 14603.8 | 14783.6 | 0.0014 | 0.0098 | 0.0000 | 0.0000 |
| M-P-SVAR | -7259.1 | 14596.1 | 14808.6 | 0.0180 | 0.0087 | 0.0000 | 0.0000 |
| M-N-C | -7378.5 | 14787.0 | 14868.7 | 0.0000 | 0.0000 | 0.0000 | 0.0000 |
| M-N-S | -7264.0 | 14593.9 | 14773.7 | 0.0000 | 0.0000 | 0.0004 | 0.1088 |
| M-N-SAR | -7173.1 | 14424.2 | 14636.6 | 0.0017 | 0.0049 | 0.0020 | 0.2389 |
| M-N-SVAR | -7161.6 | 14413.2 | 14658.4 | 0.0255 | 0.0038 | 0.0303 | 0.2009 |
| MS-P-C | -7434.2 | 14898.4 | 14980.1 | 0.0026 | 0.0000 | 0.0000 | 0.0000 |
| MS-P-S | -7320.3 | 14706.6 | 14886.4 | 0.0020 | 0.0012 | 0.0000 | 0.0000 |
| MS-P-SAR | -7264.9 | 14607.9 | 14820.3 | 0.0428 | 0.0191 | 0.0000 | 0.0000 |
| MS-P-SVAR | -7247.7 | 14585.4 | 14830.6 | 0.0260 | 0.0463 | 0.0000 | 0.0000 |
| MS-N-C | -7263.8 | 14569.6 | 14684.0 | 0.0004 | 0.0016 | 0.0000 | 0.0141 |
| MS-N-S | -7207.3 | 14492.5 | 14705.0 | 0.0008 | 0.0279 | 0.0023 | 0.1087 |
| MS-N-SAR | -7159.2 | 14408.3 | 14653.4 | 0.4061 | 0.4425 | 0.0000 | 0.0312 |
| MS-N-SVAR | -7156.7 | 14415.3 | 14693.1 | 0.6631 | 0.4248 | 0.0000 | 0.0644 |

**4 Regime specification**

| Model | $\ln \mathcal{L}_I(\theta)$ | AIC | BIC | Buys | Sells | Buys | Sells |
|---|---|---|---|---|---|---|---|
| M-P-C | -7441.0 | 14906.1 | 14971.5 | 0.0000 | 0.0000 | 0.0000 | 0.0000 |
| M-P-S | -7309.1 | 14690.3 | 14886.4 | 0.0000 | 0.0000 | 0.0000 | 0.0000 |
| M-P-SAR | -7202.9 | 14493.8 | 14733.5 | 0.0008 | 0.0076 | 0.0006 | 0.0000 |
| M-P-SVAR | -7194.4 | 14492.8 | 14776.1 | 0.0023 | 0.0084 | 0.0154 | 0.0000 |
| M-N-C | -8145.8 | 16331.6 | 16440.6 | 0.0000 | 0.0000 | 0.0000 | 0.0000 |
| M-N-S | -7256.8 | 14601.5 | 14841.2 | 0.0000 | 0.0000 | 0.0000 | 0.0076 |
| M-N-SAR | -7166.8 | 14437.6 | 14720.9 | 0.0019 | 0.0031 | 0.0031 | 0.0175 |
| M-N-SVAR | -7157.0 | 14434.1 | 14760.9 | 0.0301 | 0.0018 | 0.0000 | 0.0002 |
| MS-P-C | -7307.1 | 14662.2 | 14793.0 | 0.0907 | 0.0000 | 0.0000 | 0.0000 |
| MS-P-S | -7203.9 | 14503.8 | 14765.3 | 0.0629 | 0.0598 | 0.0061 | 0.0000 |
| MS-P-SAR | -7192.2 | 14496.4 | 14801.4 | 0.0057 | 0.0134 | 0.0000 | 0.0092 |
| MS-P-SVAR | -7194.4 | 14516.7 | 14865.3 | 0.0326 | 0.0157 | 0.0000 | 0.0000 |
| MS-N-C | -7239.7 | 14543.4 | 14717.8 | 0.0040 | 0.0988 | 0.0000 | 0.0027 |
| MS-N-S | -7158.7 | 14429.3 | 14734.4 | 0.0262 | 0.3472 | 0.0059 | 0.0008 |
| MS-N-SAR | -7161.7 | 14451.4 | 14800.0 | 0.0177 | 0.0061 | 0.0000 | 0.0033 |
| MS-N-SVAR | -7127.4 | 14398.9 | 14791.1 | 0.1008 | 0.1392 | 0.0000 | 0.0244 |

$\chi^2_{LB}$ is computed for 50 lags. $\chi^2_{CM}$ is computed for 12 cells, (0,..., 10, 11+).
$p(\chi^2)$ denotes the p-values of the corresponding test statistics. M = Static
mixture model, MS = Markov switching model, P = Poisson model, N = Negbin 2 model, C = Constants only model, S = Seasonal model, SAR = Seasonal
autoregressive model, SVAR = Seasonal vector autoregressive model.

**Table 5.6.** Model selection IBM

| Model | $\ln \mathcal{L}_I(\theta)$ | AIC | BIC | $p(\chi^2_{LB})$ | | $p(\chi^2_{CM})$ | |
|---|---|---|---|---|---|---|---|
| | | | | Buys | Sells | Buys | Sells |

| | | | | | | | |
|---|---|---|---|---|---|---|---|
| **2 Regime specification** | | | | | | | |
| M-P-C | -10356.4 | 20724.8 | 20757.5 | 0.0000 | 0.0000 | 0.0000 | 0.0000 |
| M-P-S | -9573.3 | 19182.6 | 19280.7 | 0.0000 | 0.0000 | 0.0000 | 0.0000 |
| M-P-SAR | -9089.4 | 18222.8 | 18342.6 | 0.0000 | 0.0000 | 0.0000 | 0.0000 |
| M-P-SVAR | -9019.9 | 18091.9 | 18233.5 | 0.0000 | 0.0000 | 0.0000 | 0.0000 |
| M-N-C | -8863.0 | 17746.0 | 17800.5 | 0.0000 | 0.0000 | 0.3819 | 0.6301 |
| M-N-S | -8609.1 | 17262.1 | 17382.0 | 0.0000 | 0.0000 | 0.0029 | 0.1764 |
| M-N-SAR | -8407.8 | 16867.6 | 17009.2 | 0.0000 | 0.0000 | 0.0065 | 0.0040 |
| M-N-SVAR | -8376.8 | 16813.6 | 16977.0 | 0.0000 | 0.0000 | 0.0001 | 0.0128 |
| MS-P-C | -10216.3 | 20448.7 | 20492.3 | 0.0000 | 0.0000 | 0.0000 | 0.0000 |
| MS-P-S | -9519.7 | 19079.5 | 19188.4 | 0.0000 | 0.0000 | 0.0000 | 0.0000 |
| MS-P-SAR | -9088.9 | 18225.8 | 18356.6 | 0.0000 | 0.0000 | 0.0000 | 0.0000 |
| MS-P-SVAR | -9020.0 | 18095.9 | 18248.5 | 0.0000 | 0.0000 | 0.0000 | 0.0000 |
| MS-N-C | -8640.5 | 17304.9 | 17370.3 | 0.0000 | 0.0000 | 0.0000 | 0.0101 |
| MS-N-S | -8540.9 | 17129.8 | 17260.6 | 0.0000 | 0.0000 | 0.0164 | 0.0939 |
| MS-N-SAR | -8407.9 | 16871.8 | 17024.3 | 0.0000 | 0.0000 | 0.0070 | 0.0042 |
| MS-N-SVAR | -8376.2 | 16816.3 | 16990.7 | 0.0000 | 0.0000 | 0.0000 | 0.0061 |
| **3 Regime specification** | | | | | | | |
| M-P-C | -9549.1 | 19116.3 | 19165.3 | 0.0000 | 0.0000 | 0.0000 | 0.0000 |
| M-P-S | -9041.7 | 18137.4 | 18284.5 | 0.0000 | 0.0000 | 0.0000 | 0.0000 |
| M-P-SAR | -8699.4 | 17464.8 | 17644.5 | 0.0000 | 0.0000 | 0.0000 | 0.0000 |
| M-P-SVAR | -8640.6 | 17359.2 | 17571.7 | 0.0000 | 0.0000 | 0.0000 | 0.0000 |
| M-N-C | -8862.8 | 17755.7 | 17837.4 | 0.0000 | 0.0000 | 0.0024 | 0.0000 |
| M-N-S | -8590.4 | 17246.8 | 17426.6 | 0.0000 | 0.0000 | 0.0308 | 0.3748 |
| M-N-SAR | -8374.7 | 16827.5 | 17039.9 | 0.0000 | 0.0000 | 0.5241 | 0.0308 |
| M-N-SVAR | -8344.9 | 16779.8 | 17024.9 | 0.0000 | 0.0000 | 0.0194 | 0.0338 |
| MS-P-C | -9260.5 | 18551.0 | 18632.7 | 0.0000 | 0.0000 | 0.0000 | 0.0000 |
| MS-P-S | -8866.0 | 17798.1 | 17977.9 | 0.0000 | 0.0000 | 0.0000 | 0.0000 |
| MS-P-SAR | -8697.6 | 17473.2 | 17685.7 | 0.0000 | 0.0000 | 0.0000 | 0.0000 |
| MS-P-SVAR | -8637.9 | 17365.9 | 17611.0 | 0.0000 | 0.0000 | 0.0000 | 0.0000 |
| MS-N-C | -8502.0 | 17045.9 | 17160.3 | 0.0000 | 0.0000 | 0.0000 | 0.1177 |
| MS-N-S | -8364.3 | 16806.7 | 17019.1 | 0.0000 | 0.0000 | 0.0033 | 0.0000 |
| MS-N-SAR | -8297.1 | 16684.3 | 16929.4 | 0.0054 | 0.0121 | 0.0001 | 0.0000 |
| MS-N-SVAR | -8285.7 | 16673.4 | 16951.2 | 0.0092 | 0.0112 | 0.0000 | 0.0000 |
| **4 Regime specification** | | | | | | | |
| M-P-C | -9210.3 | 18444.7 | 18510.0 | 0.0000 | 0.0000 | 0.0000 | 0.0000 |
| M-P-S | -8784.3 | 17640.6 | 17836.7 | 0.0000 | 0.0000 | 0.0000 | 0.0000 |
| M-P-SAR | -8581.9 | 17251.9 | 17491.6 | 0.0000 | 0.0000 | 0.0000 | 0.0000 |
| M-P-SVAR | -8505.4 | 17114.9 | 17398.1 | 0.0000 | 0.0000 | 0.0000 | 0.0000 |
| M-N-C | -8864.7 | 17769.4 | 17878.3 | 0.0000 | 0.0000 | 0.0000 | 0.0000 |
| M-N-S | -8609.7 | 17307.5 | 17547.2 | 0.0000 | 0.0000 | 0.0000 | 0.0000 |
| M-N-SAR | -8373.0 | 16849.9 | 17133.2 | 0.0000 | 0.0000 | 0.0000 | 0.0000 |
| M-N-SVAR | -8318.9 | 16757.9 | 17084.7 | 0.0000 | 0.0000 | 0.0261 | 0.0706 |
| MS-P-C | -8887.0 | 17822.0 | 17952.8 | 0.0000 | 0.0000 | 0.0000 | 0.0000 |
| MS-P-S | -8541.0 | 17178.0 | 17439.5 | 0.0000 | 0.0000 | 0.0000 | 0.0000 |
| MS-P-SAR | -8455.1 | 17022.3 | 17327.3 | 0.0001 | 0.0000 | 0.0000 | 0.0000 |
| MS-P-SVAR | -8442.8 | 17013.6 | 17362.2 | 0.0003 | 0.0000 | 0.0000 | 0.0000 |
| MS-N-C | -8448.6 | 16961.2 | 17135.5 | 0.0000 | 0.0000 | 0.0027 | 0.0058 |
| MS-N-S | -8345.9 | 16803.8 | 17108.9 | 0.0000 | 0.0000 | 0.0000 | 0.0019 |
| MS-N-SAR | -8306.0 | 16739.9 | 17088.6 | 0.0002 | 0.0000 | 0.0353 | 0.0421 |
| MS-N-SVAR | -8226.7 | 16597.4 | 16989.5 | 0.0657 | 0.0109 | 0.0001 | 0.0232 |

$\chi^2_{LB}$ is computed for 50 lags. $\chi^2_{CM}$ is computed for 12 cells, (0,..., 10, 11+).
$p(\chi^2)$ denotes the p-values of the corresponding test statistics. M = Static mixture model, MS = Markov switching model, P = Poisson model, N = Negbin 2 model, C = Constants only model, S = Seasonal model, SAR = Seasonal autoregressive model, SVAR = Seasonal vector autoregressive model.

**Table 5.7.** Model selection KO

| Model | $\ln \mathcal{L}_I(\theta)$ | $AIC$ | $BIC$ | $p(\chi^2_{LB})$ | | $p(\chi^2_{CM})$ | |
|---|---|---|---|---|---|---|---|
| | | | | Buys | Sells | Buys | Sells |
| | *2 Regime specification* | | | | | | |
| M-P-C | -8368.2 | 16748.4 | 16781.1 | 0.0000 | 0.0000 | 0.0000 | 0.0000 |
| M-P-S | -8131.3 | 16298.6 | 16396.6 | 0.0000 | 0.0000 | 0.0000 | 0.0000 |
| M-P-SAR | -7979.7 | 16003.4 | 16123.2 | 0.6191 | 0.2057 | 0.0000 | 0.0000 |
| M-P-SVAR | -7976.2 | 16004.5 | 16146.1 | 0.6156 | 0.2366 | 0.0000 | 0.0000 |
| M-N-C | -7942.8 | 15905.6 | 15960.1 | 0.0000 | 0.0000 | 0.0000 | 0.2756 |
| M-N-S | -7818.4 | 15680.7 | 15800.6 | 0.0000 | 0.0000 | 0.0000 | 0.3088 |
| M-N-SAR | -7730.6 | 15513.2 | 15654.9 | 0.6250 | 0.1678 | 0.0000 | 0.3052 |
| M-N-SVAR | -7724.3 | 15508.6 | 15672.0 | 0.6458 | 0.2083 | 0.0000 | 0.2170 |
| MS-P-C | -8340.2 | 16696.4 | 16739.9 | 0.0000 | 0.0000 | 0.0000 | 0.0000 |
| MS-P-S | -8102.6 | 16245.2 | 16354.1 | 0.0061 | 0.0000 | 0.0000 | 0.0000 |
| MS-P-SAR | -7979.2 | 16006.4 | 16137.1 | 0.6488 | 0.2072 | 0.0000 | 0.0000 |
| MS-P-SVAR | -7976.2 | 16008.3 | 16160.9 | 0.6280 | 0.2378 | 0.0000 | 0.0000 |
| MS-N-C | -7912.8 | 15849.6 | 15914.9 | 0.0000 | 0.0000 | 0.0007 | 0.2280 |
| MS-N-S | -7779.6 | 15607.2 | 15738.0 | 0.0006 | 0.0000 | 0.0001 | 0.1169 |
| MS-N-SAR | -7728.2 | 15512.5 | 15665.0 | 0.6155 | 0.1532 | 0.0000 | 0.1181 |
| MS-N-SVAR | -7723.1 | 15510.1 | 15684.4 | 0.6223 | 0.2052 | 0.0000 | 0.0748 |
| | *3 Regime specification* | | | | | | |
| M-P-C | -8087.0 | 16192.0 | 16241.0 | 0.0000 | 0.0000 | 0.0000 | 0.0000 |
| M-P-S | -7922.7 | 15899.4 | 16046.5 | 0.0000 | 0.0000 | 0.0000 | 0.0000 |
| M-P-SAR | -7804.5 | 15675.0 | 15854.8 | 0.6525 | 0.1615 | 0.0000 | 0.0000 |
| M-P-SVAR | -7799.8 | 15677.6 | 15890.0 | 0.3444 | 0.2611 | 0.0000 | 0.0000 |
| M-N-C | -7940.1 | 15910.2 | 15991.9 | 0.0000 | 0.0000 | 0.0000 | 0.0262 |
| M-N-S | -7804.0 | 15674.0 | 15853.8 | 0.0000 | 0.0000 | 0.0126 | 0.0475 |
| M-N-SAR | -7716.1 | 15510.1 | 15722.6 | 0.6021 | 0.1631 | 0.4434 | 0.1374 |
| M-N-SVAR | -7693.9 | 15477.8 | 15722.9 | 0.7388 | 0.2208 | 0.0119 | 0.1240 |
| MS-P-C | -7991.7 | 16013.5 | 16095.2 | 0.0114 | 0.0002 | 0.0000 | 0.0000 |
| MS-P-S | -7847.5 | 15761.0 | 15940.8 | 0.5120 | 0.0683 | 0.0000 | 0.0000 |
| MS-P-SAR | -7790.0 | 15658.1 | 15870.5 | 0.7543 | 0.3151 | 0.0000 | 0.0000 |
| MS-P-SVAR | -7776.1 | 15642.1 | 15887.2 | 0.7953 | 0.4059 | 0.0000 | 0.0000 |
| MS-N-C | -7830.4 | 15702.7 | 15817.1 | 0.0146 | 0.0000 | 0.0000 | 0.1289 |
| MS-N-S | -7730.8 | 15539.5 | 15752.0 | 0.1034 | 0.0108 | 0.0058 | 0.0001 |
| MS-N-SAR | -7693.6 | 15477.3 | 15722.4 | 0.7559 | 0.4065 | 0.0590 | 0.3424 |
| MS-N-SVAR | -7681.4 | 15464.7 | 15742.5 | 0.7311 | 0.5138 | 0.0676 | 0.1668 |
| | *4 Regime specification* | | | | | | |
| M-P-C | -8005.4 | 16034.7 | 16100.1 | 0.0000 | 0.0000 | 0.0000 | 0.0000 |
| M-P-S | -7864.7 | 15801.3 | 15997.5 | 0.0000 | 0.0000 | 0.0000 | 0.0000 |
| M-P-SAR | -7732.2 | 15552.4 | 15792.1 | 0.6723 | 0.1858 | 0.0000 | 0.0046 |
| M-P-SVAR | -7732.6 | 15569.2 | 15852.4 | 0.7778 | 0.2998 | 0.0000 | 0.0047 |
| M-N-C | -7932.6 | 15905.2 | 16014.1 | 0.0000 | 0.0000 | 0.0001 | 0.3948 |
| M-N-S | -7789.6 | 15667.2 | 15906.9 | 0.0000 | 0.0000 | 0.0017 | 0.0195 |
| M-N-SAR | -7703.9 | 15511.9 | 15795.1 | 0.6636 | 0.1797 | 0.1797 | 0.0332 |
| M-N-SVAR | -7685.7 | 15491.5 | 15818.3 | 0.7373 | 0.1528 | 0.0148 | 0.0244 |
| MS-P-C | -7893.0 | 15834.1 | 15964.8 | 0.0896 | 0.0003 | 0.0000 | 0.0000 |
| MS-P-S | -7750.0 | 15596.1 | 15857.6 | 0.7195 | 0.1086 | 0.0000 | 0.0000 |
| MS-P-SAR | -7707.7 | 15527.4 | 15832.4 | 0.7657 | 0.2380 | 0.0005 | 0.0003 |
| MS-P-SVAR | -7711.9 | 15551.7 | 15900.4 | 0.7668 | 0.3667 | 0.0000 | 0.0053 |
| MS-N-C | -7784.1 | 15632.2 | 15806.5 | 0.0697 | 0.1364 | 0.0646 | 0.0444 |
| MS-N-S | -7685.6 | 15483.1 | 15788.3 | 0.3374 | 0.2540 | 0.0229 | 0.0650 |
| MS-N-SAR | -7685.2 | 15498.4 | 15847.0 | 0.7675 | 0.3495 | 0.0112 | 0.2193 |
| MS-N-SVAR | -7666.7 | 15477.5 | 15869.7 | 0.7688 | 0.4437 | 0.0335 | 0.2656 |

$\chi^2_{LB}$ is computed for 50 lags. $\chi^2_{CM}$ is computed for 12 cells, $(0,..., 10, 11+)$.
$p(\chi^2)$ denotes the p-values of the corresponding test statistics. M = Static mixture model, MS = Markov switching model, P = Poisson model, N = Negbin 2 model, C = Constants only model, S = Seasonal model, SAR = Seasonal autoregressive model, SVAR = Seasonal vector autoregressive model.

**Table 5.8.** Model selection XON

| Model | $\ln \mathcal{L}_I(\theta)$ | $AIC$ | $BIC$ | $p(\chi^2_{LB})$ | | $p(\chi^2_{CM})$ | |
|---|---|---|---|---|---|---|---|
| | | | | Buys | Sells | Buys | Sells |

| Model | $\ln \mathcal{L}_I(\theta)$ | $AIC$ | $BIC$ | Buys | Sells | Buys | Sells |
|---|---|---|---|---|---|---|---|
| **2 Regime specification** | | | | | | | |
| M-P-C | -7833.9 | 15679.7 | 15712.4 | 0.0000 | 0.0000 | 0.0000 | 0.0000 |
| M-P-S | -7615.5 | 15267.0 | 15365.1 | 0.0000 | 0.0000 | 0.0000 | 0.0000 |
| M-P-SAR | -7476.4 | 14996.7 | 15116.5 | 0.0000 | 0.0001 | 0.0000 | 0.0000 |
| M-P-SVAR | -7463.7 | 14979.4 | 15121.0 | 0.0000 | 0.0001 | 0.0000 | 0.0000 |
| M-N-C | -7383.4 | 14786.7 | 14841.2 | 0.0000 | 0.0000 | 0.0000 | 0.1343 |
| M-N-S | -7289.0 | 14622.0 | 14741.9 | 0.0000 | 0.0000 | 0.0000 | 0.1320 |
| M-N-SAR | -7195.0 | 14442.0 | 14583.7 | 0.0000 | 0.0002 | 0.0000 | 0.1280 |
| M-N-SVAR | -7185.3 | 14430.6 | 14594.0 | 0.0000 | 0.0002 | 0.0000 | 0.1814 |
| MS-P-C | -7780.3 | 15576.5 | 15620.1 | 0.0000 | 0.0000 | 0.0000 | 0.0000 |
| MS-P-S | -7563.3 | 15166.7 | 15275.6 | 0.0000 | 0.0000 | 0.0000 | 0.0000 |
| MS-P-SAR | -7473.1 | 14994.1 | 15124.8 | 0.0000 | 0.0001 | 0.0000 | 0.0000 |
| MS-P-SVAR | -7456.4 | 14968.8 | 15121.3 | 0.0000 | 0.0001 | 0.0000 | 0.0000 |
| MS-N-C | -7321.3 | 14667.2 | 14732.6 | 0.0000 | 0.0000 | 0.0927 | 0.2295 |
| MS-N-S | -7234.0 | 14516.0 | 14646.7 | 0.0000 | 0.0000 | 0.0093 | 0.1666 |
| MS-N-SAR | -7187.5 | 14431.1 | 14583.6 | 0.0000 | 0.0003 | 0.0494 | 0.0115 |
| MS-N-SVAR | -7182.1 | 14428.2 | 14602.5 | 0.0000 | 0.0005 | 0.0028 | 0.0626 |
| **3 Regime specification** | | | | | | | |
| M-P-C | -7578.1 | 15174.3 | 15223.9 | 0.0000 | 0.0000 | 0.0000 | 0.0000 |
| M-P-S | -7410.5 | 14874.9 | 15022.0 | 0.0000 | 0.0000 | 0.0000 | 0.0000 |
| M-P-SAR | -7295.8 | 14657.7 | 14837.4 | 0.0000 | 0.0004 | 0.0000 | 0.0000 |
| M-P-SVAR | -7300.6 | 14679.2 | 14891.7 | 0.0000 | 0.0018 | 0.0000 | 0.0000 |
| M-N-C | -7383.4 | 14796.7 | 14878.5 | 0.0000 | 0.0000 | 0.0000 | 0.0000 |
| M-N-S | -7257.2 | 14580.4 | 14760.2 | 0.0000 | 0.0000 | 0.0000 | 0.1258 |
| M-N-SAR | -7164.6 | 14407.3 | 14619.7 | 0.0000 | 0.0017 | 0.0000 | 0.0477 |
| M-N-SVAR | -7225.5 | 14541.0 | 14786.1 | 0.0000 | 0.0004 | 0.0000 | 0.0000 |
| MS-P-C | -7491.6 | 15013.2 | 15094.9 | 0.0000 | 0.0000 | 0.0000 | 0.0000 |
| MS-P-S | -7338.9 | 14743.8 | 14923.5 | 0.0000 | 0.0000 | 0.0000 | 0.0000 |
| MS-P-SAR | -7287.5 | 14653.0 | 14865.4 | 0.0000 | 0.0014 | 0.0000 | 0.0000 |
| MS-P-SVAR | -7282.5 | 14655.0 | 14900.1 | 0.0000 | 0.0193 | 0.0000 | 0.0000 |
| MS-N-C | -7248.6 | 14539.2 | 14653.6 | 0.0069 | 0.0004 | 0.0000 | 0.1125 |
| MS-N-S | -7162.9 | 14403.8 | 14616.3 | 0.0000 | 0.0195 | 0.0000 | 0.0003 |
| MS-N-SAR | -7131.5 | 14352.9 | 14598.1 | 0.0087 | 0.1402 | 0.0007 | 0.0055 |
| MS-N-SVAR | -7120.7 | 14343.3 | 14621.2 | 0.0130 | 0.0642 | 0.0004 | 0.0029 |
| **4 Regime specification** | | | | | | | |
| M-P-C | -7472.7 | 14969.4 | 15034.8 | 0.0000 | 0.0000 | 0.0000 | 0.0000 |
| M-P-S | -7308.4 | 14688.9 | 14885.0 | 0.0000 | 0.0000 | 0.0000 | 0.0000 |
| M-P-SAR | -7203.8 | 14495.6 | 14735.3 | 0.0000 | 0.0006 | 0.0000 | 0.0000 |
| M-P-SVAR | -7224.8 | 14553.7 | 14836.9 | 0.0000 | 0.0018 | 0.0000 | 0.0000 |
| M-N-C | -7383.0 | 14806.1 | 14915.0 | 0.0000 | 0.0000 | 0.0000 | 0.0000 |
| M-N-S | -7265.9 | 14619.8 | 14859.5 | 0.0000 | 0.0000 | 0.0000 | 0.0000 |
| M-N-SAR | -7172.8 | 14449.6 | 14732.8 | 0.0000 | 0.0009 | 0.0009 | 0.0000 |
| M-N-SVAR | -7204.7 | 14529.4 | 14856.3 | 0.0000 | 0.0004 | 0.0000 | 0.0000 |
| MS-P-C | -7328.7 | 14705.4 | 14836.1 | 0.0004 | 0.0000 | 0.0000 | 0.0000 |
| MS-P-S | -7177.5 | 14451.0 | 14712.5 | 0.0000 | 0.0399 | 0.0005 | 0.0000 |
| MS-P-SAR | -7159.3 | 14430.6 | 14735.6 | 0.0000 | 0.0101 | 0.0068 | 0.0000 |
| MS-P-SVAR | -7170.3 | 14468.5 | 14817.1 | 0.0000 | 0.0108 | 0.0000 | 0.0000 |
| MS-N-C | -7214.6 | 14493.2 | 14667.5 | 0.0689 | 0.0187 | 0.0000 | 0.0151 |
| MS-N-S | -7123.0 | 14358.1 | 14663.1 | 0.2533 | 0.0321 | 0.0001 | 0.0002 |
| MS-N-SAR | -7104.7 | 14337.4 | 14686.1 | 0.5951 | 0.7014 | 0.0000 | 0.0009 |
| MS-N-SVAR | -7151.8 | 14447.6 | 14839.8 | 0.0000 | 0.3075 | 0.0000 | 0.0000 |

$\chi^2_{LB}$ is computed for 50 lags. $\chi^2_{CM}$ is computed for 12 cells, (0,..., 10, 11+).
$p(\chi^2)$ denotes the p-values of the corresponding test statistics. M = Static mixture model, MS = Markov switching model, P = Poisson model, N = Negbin 2 model, C = Constants only model, S = Seasonal model, SAR = Seasonal autoregressive model, SVAR = Seasonal vector autoregressive model.

## 5.4.2 Parameter Estimates

The first step in our estimation strategy involved an assessment of the model specification that appeared to be appropriate for our sample data. We found, that for all stocks in our sample the negative binomial 3 regime Markov switching regression model with an autoregressive conditional mean specification accompanied by seasonal components performed best, when judged by the value of the $BIC$. Tables 5.9 through 5.13 contain the parameter estimates of the preferred model specifications for the stocks in our sample, along with standard errors and $t$-values, computed from the QML-estimator of the information matrix associated with the expected complete log-likelihood function $\ln \mathcal{L}_{EC}(\theta)$ and results of the specification tests we conducted. Instead of discussing the economic implications for every single parameter estimate in detail, we want to highlight some of the apparent features of our estimates and point out some similarities and differences between the stocks in our sample.

First, we find that all of the $\alpha$ parameters, except for the XON buys under regime 1, are significantly different from zero as indicated by the associated $t$-values. Furthermore, we find them all to be less than one in value, thus implying that the regime specific overdispersion relative to a Poisson density is modest. We interpret this finding as evidence in favor of the existence of heterogeneity of the trading behavior not only between the two groups of informed and uninformed traders, but also within each group, i.e. not all informed traders tend to exploit their informational advantage in the same way. If this result was caused only by the behavior of informed traders, then we would expect insignificant estimates of the parameter $\alpha$ under the no news regime, but not under the bad and the good regime, as the trading population under the no news regime should consist only of uninformed traders. This is however not the case, as we find significant estimates of $\alpha$ under all three regimes. In addition, we find that the composition of the trading population is different not only across regimes, but also across equations, thus implying that there is heterogeneity among buyers and sellers irrespective of their level of information.

Second, we observe that the autoregressive parameters $\phi_B$ and $\phi_S$ are generally positive and significant across all regimes and equations. The estimates of the autoregressive parameters typically lie in the range between 0.1 and 0.5, with only three exceptions where we find higher values between 0.719 and 0.941 for $\phi_B$ under regime 1 (BA, IBM, XON). Note, that none of the parameters is larger than one in value, implying that the regime specific trading processes are stationary in general. However, in a few cases we find negative values, but these tend to be small in magnitude and are only significantly different from zero in one case (XON sells under regime 1). These results underscore that there appears to be some momentum present in the trading process in the sense that trades on one side of the market tend to trigger trades on the same side of the market, irrespective of the nature of the present information regime. This trading momentum appears to be moderate in size, as the estimates of the autoregressive parameters tend to be generally small.

A third finding, probably of minor importance, is that the estimates of the parameter $\delta$, corresponding to the first 5 minute interval of each trading day is generally negative and significantly different from zero, with the exception of the third regime, where we find it to be insignificant for both equations and for all stocks, except for KO and XON. This seems to indicate, that when the information regime is in state three, the active trading population prefers to avoid trading during the opening batch auction and perhaps a short time period afterwards. This evidence must be interpreted cautiously however, since the transactions that are conducted during the opening batch auctions are not included in our data set, and thus the result might as well be reversed, if such trades were recorded and could be classified as buyer or seller initiated.

Finally, we find that the estimates of the transition probabilities $p_{ji}$ tend to have large values for the probabilities of staying in the same regime as before, i.e. $p_{jj}$, and for the transitions between regime 1 and regime 2, i.e. $p_{21}$ and $p_{12}$. Estimates of the transition probabilities between regimes 1 and 2 on the one hand and the third regime on the other hand tend to be low, often with values less than 1%. This appears to be related to the high persistence of the third regime making transitions between this regime and the remaining two a rare event in practice. We observe that the probability of staying in regime 3, $p_{33}$, has the highest estimated values for all stocks in our sample. Recalling, that the duration of staying in a given regime once it is entered is completely determined by the magnitude of the diagonal entries $p_{jj}$ in the transition matrix $P$, this finding indicates that for the stocks in our sample the third regime appears to be the most persistent one. When the information process leaves regime 3, it appears to be frequently switching between regimes 1 and 2. The associated informational epochs appear to be short-lived, when judged by the magnitudes of the diagonal elements $p_{11}$ and $p_{22}$.

In Tables 5.9 through 5.13 we additionally include $t$-values for the Null hypothesis, that the estimates of the transition probabilities from state $i$ into state $j$ are equal to the unconditional (ergodic) probability of being in state $j$, i.e. $H_0 : p_{ji} = \pi_j$. These tend to be insignificant at the 10% significance level in general, with only three exceptions for the probabilities $p_{21}$ and $p_{12}$ of the KO stock and $p_{21}$ for XON, who are fairly close to the corresponding unconditional probabilities. Noting, that in the special case, when all transition probabilities are equal to the unconditional probabilities $p_{ji} = \pi_j$, the Markov switching model would essentially be equal to a static mixture model, we interpret the rejection of the Null hypothesis as a confirmation in favor of using the Markov switching model rather than a static mixture model to describe the trading process for the stocks in our sample.

In order to get an impression of how well the 3-regime Markov switching regression model describes our time series of trading events, we plot the unconditional forecasts for each variable against the sample realization for an arbitrarily chosen period of three trading days (August 19-21, 1996) in Fig.

5.4. Note, that episodes of heavy trading appear to be concentrated at the beginning and the end of each trading day, thus contributing to the U-shaped intradaily seasonal pattern that we found to be characteristic for our sample time series. During the rest of the day trading activity appears to be moderately compared to the picture near the open and the close. When there are episodes of heavy trading activity during the middle part of the day they appear to be clustered, thus contributing to the pattern of serial dependence. Note, that the forecasts smooth through these episodes of heavy trading, generally following the specified sinusoidal seasonal pattern, but abandoning the seasonal path in quick response to any exceptional activity that occurs. In general, the movements of the forecasts are in line with the direction of trading activity. It is also apparent, that large deviations from the expected seasonal level of trading are not completely matched by the forecasts. The magnitudes of these eruptions of activity seem to be very difficult to forecast.

**Table 5.9.** Parameter estimates BA

| Parameter | Equation | Estimate | Std.-error | $t$-statistic |
|---|---|---|---|---|
| $\omega^{(1)}$ | $B_t$ | -1.051 | 0.078 | -13.543 |
| $\zeta^{(1)}$ | $B_t$ | 0.049 | 0.065 | 0.759 |
| $\eta^{(1)}$ | $B_t$ | 0.031 | 0.063 | 0.490 |
| $\delta^{(1)}$ | $B_t$ | -13.837 | 0.468 | -29.575 |
| $\phi_B^{(1)}$ | $B_t$ | 0.941 | 0.048 | 19.552 |
| $\alpha^{(1)}$ | $B_t$ | 0.198 | 0.065 | 3.063 |
| $\omega^{(1)}$ | $S_t$ | 1.414 | 0.031 | 45.241 |
| $\zeta^{(1)}$ | $S_t$ | 0.026 | 0.037 | 0.700 |
| $\eta^{(1)}$ | $S_t$ | 0.129 | 0.039 | 3.324 |
| $\delta^{(1)}$ | $S_t$ | -0.690 | 0.273 | -2.530 |
| $\phi_S^{(1)}$ | $S_t$ | -0.014 | 0.026 | -0.538 |
| $\alpha^{(1)}$ | $S_t$ | 0.073 | 0.021 | 3.460 |
| $\omega^{(2)}$ | $B_t$ | 1.276 | 0.037 | 34.098 |
| $\zeta^{(2)}$ | $B_t$ | 0.041 | 0.040 | 1.022 |
| $\eta^{(2)}$ | $B_t$ | 0.159 | 0.042 | 3.805 |
| $\delta^{(2)}$ | $B_t$ | -0.311 | 0.300 | -1.036 |
| $\phi_B^{(2)}$ | $B_t$ | 0.110 | 0.030 | 3.670 |
| $\alpha^{(2)}$ | $B_t$ | 0.416 | 0.035 | 12.040 |
| $\omega^{(2)}$ | $S_t$ | 0.226 | 0.044 | 5.107 |
| $\zeta^{(2)}$ | $S_t$ | -0.027 | 0.051 | -0.524 |
| $\eta^{(2)}$ | $S_t$ | 0.209 | 0.051 | 4.110 |
| $\delta^{(2)}$ | $S_t$ | -0.736 | 0.296 | -2.488 |
| $\phi_S^{(2)}$ | $S_t$ | 0.418 | 0.036 | 11.485 |
| $\alpha^{(2)}$ | $S_t$ | 0.569 | 0.066 | 8.663 |
| $\omega^{(3)}$ | $B_t$ | 0.422 | 0.037 | 11.271 |
| $\zeta^{(3)}$ | $B_t$ | 0.068 | 0.048 | 1.402 |
| $\eta^{(3)}$ | $B_t$ | 0.253 | 0.048 | 5.256 |
| $\delta^{(3)}$ | $B_t$ | -0.441 | 0.353 | -1.250 |
| $\phi_B^{(3)}$ | $B_t$ | 0.225 | 0.040 | 5.639 |
| $\alpha^{(3)}$ | $B_t$ | 0.499 | 0.058 | 8.615 |
| $\omega^{(3)}$ | $S_t$ | 0.263 | 0.038 | 6.903 |
| $\zeta^{(3)}$ | $S_t$ | 0.076 | 0.049 | 1.544 |
| $\eta^{(3)}$ | $S_t$ | 0.158 | 0.053 | 3.007 |
| $\delta^{(3)}$ | $S_t$ | -0.221 | 0.260 | -0.848 |
| $\phi_S^{(3)}$ | $S_t$ | 0.201 | 0.043 | 4.688 |
| $\alpha^{(3)}$ | $S_t$ | 0.489 | 0.063 | 7.759 |
| $p_{11}$ | - | 0.2785 | 0.014 | 7.765 |
| $p_{21}$ | - | 0.6563 | 0.015 | 18.531 |
| $p_{31}$ | - | 0.0652 | - | - |
| $p_{12}$ | - | 0.3017 | 0.011 | 11.802 |
| $p_{22}$ | - | 0.6981 | 0.011 | 28.131 |
| $p_{32}$ | - | 0.0002 | - | - |
| $p_{13}$ | - | 0.0152 | 0.001 | -115.897 |
| $p_{23}$ | - | 0.0101 | 0.001 | -486.720 |
| $p_{33}$ | - | 0.9747 | - | - |

| | | | | |
|---|---|---|---|---|
| $\ln \mathcal{L}_I(\theta)$ | -6545.9 | $\chi^2_{LB}$ | $B_t$ | 64.0 |
| $T$ | 1715 | | $S_t$ | 53.7 |
| $AIC$ | 13181.7 | $\chi^2_{CM}$ | $B_t$ | 36.2 |
| $BIC$ | 13426.9 | | $S_t$ | 16.0 |

$\chi^2_{LB}$ is computed for 50 lags. The corresponding tabulated 99% (95%) critical value for 50 degrees of freedom is 76.2 (67.5). $\chi^2_{CM}$ is computed for 12 cells, (0,..., 10, 11+). The corresponding tabulated 99% (95%) critical value for 11 degrees of freedom is 24.7 (19.7). The $t$-statistic for the regime probabilities corresponds to $H_0$ : $p_{ji} = \pi_j$.

**Table 5.10.** Parameter estimates DIS

| Parameter | Equation | Estimate | Std.-error | $t$-statistic |
|---|---|---|---|---|
| $\omega^{(1)}$ | $B_t$ | -0.105 | 0.066 | -1.596 |
| $\zeta^{(1)}$ | $B_t$ | 0.204 | 0.056 | 3.666 |
| $\eta^{(1)}$ | $B_t$ | 0.059 | 0.057 | 1.040 |
| $\delta^{(1)}$ | $B_t$ | 1.152 | 0.150 | 7.662 |
| $\phi_B^{(1)}$ | $B_t$ | 0.500 | 0.042 | 11.818 |
| $\alpha^{(1)}$ | $B_t$ | 0.217 | 0.055 | 3.951 |
| $\omega^{(1)}$ | $S_t$ | 1.323 | 0.046 | 28.956 |
| $\zeta^{(1)}$ | $S_t$ | 0.033 | 0.041 | 0.810 |
| $\eta^{(1)}$ | $S_t$ | 0.239 | 0.044 | 5.458 |
| $\delta^{(1)}$ | $S_t$ | -1.426 | 0.317 | -4.503 |
| $\phi_S^{(1)}$ | $S_t$ | 0.230 | 0.034 | 6.833 |
| $\alpha^{(1)}$ | $S_t$ | 0.110 | 0.026 | 4.162 |
| $\omega^{(2)}$ | $B_t$ | 1.590 | 0.043 | 37.324 |
| $\zeta^{(2)}$ | $B_t$ | 0.018 | 0.035 | 0.520 |
| $\eta^{(2)}$ | $B_t$ | 0.309 | 0.041 | 7.597 |
| $\delta^{(2)}$ | $B_t$ | -1.177 | 0.192 | -6.125 |
| $\phi_B^{(2)}$ | $B_t$ | 0.020 | 0.033 | 0.609 |
| $\alpha^{(2)}$ | $B_t$ | 0.230 | 0.026 | 8.739 |
| $\omega^{(2)}$ | $S_t$ | 0.363 | 0.042 | 8.668 |
| $\zeta^{(2)}$ | $S_t$ | 0.070 | 0.045 | 1.539 |
| $\eta^{(2)}$ | $S_t$ | 0.561 | 0.049 | 11.380 |
| $\delta^{(2)}$ | $S_t$ | 1.003 | 0.396 | 2.531 |
| $\phi_S^{(2)}$ | $S_t$ | 0.189 | 0.036 | 5.200 |
| $\alpha^{(2)}$ | $S_t$ | 0.306 | 0.046 | 6.577 |
| $\omega^{(3)}$ | $B_t$ | 0.792 | 0.035 | 22.423 |
| $\zeta^{(3)}$ | $B_t$ | 0.046 | 0.035 | 1.299 |
| $\eta^{(3)}$ | $B_t$ | 0.155 | 0.038 | 4.031 |
| $\delta^{(3)}$ | $B_t$ | -0.234 | 0.288 | -0.811 |
| $\phi_B^{(3)}$ | $B_t$ | 0.239 | 0.030 | 7.888 |
| $\alpha^{(3)}$ | $B_t$ | 0.377 | 0.038 | 9.902 |
| $\omega^{(3)}$ | $S_t$ | 0.675 | 0.034 | 20.123 |
| $\zeta^{(3)}$ | $S_t$ | 0.145 | 0.040 | 3.633 |
| $\eta^{(3)}$ | $S_t$ | 0.145 | 0.041 | 3.514 |
| $\delta^{(3)}$ | $S_t$ | -0.076 | 0.169 | -0.449 |
| $\phi_S^{(3)}$ | $S_t$ | 0.059 | 0.034 | 1.733 |
| $\alpha^{(3)}$ | $S_t$ | 0.387 | 0.042 | 9.290 |
| $p_{11}$ | - | 0.4718 | 0.019 | 16.341 |
| $p_{21}$ | - | 0.5281 | 0.019 | 13.115 |
| $p_{31}$ | - | 0.0001 | - | - |
| $p_{12}$ | - | 0.2990 | 0.012 | 11.940 |
| $p_{22}$ | - | 0.6734 | 0.012 | 34.078 |
| $p_{32}$ | - | 0.0276 | - | - |
| $p_{13}$ | - | 0.0030 | 0.000 | -513.304 |
| $p_{23}$ | - | 0.0108 | 0.001 | -311.946 |
| $p_{33}$ | - | 0.9862 | - | - |

| $\ln \mathcal{L}_I(\theta)$ | -7159.2 | $\chi^2_{LB}$ | $B_t$ | 51.7 |
|---|---|---|---|---|
| $T$ | 1715 | | $S_t$ | 50.8 |
| $AIC$ | 14408.3 | $\chi^2_{CM}$ | $B_t$ | 41.9 |
| $BIC$ | 14653.4 | | $S_t$ | 21.2 |

$\chi^2_{LB}$ is computed for 50 lags. The corresponding tabulated 99% (95%) critical value for 50 degrees of freedom is 76.2 (67.5). $\chi^2_{CM}$ is computed for 12 cells, (0,..., 10, 11+). The corresponding tabulated 99% (95%) critical value for 11 degrees of freedom is 24.7 (19.7). The $t$-statistic for the regime probabilities corresponds to $H_0$ : $p_{ji} = \pi_j$.

**Table 5.11.** Parameter estimates IBM

| Parameter | Equation | Estimate | Std.-error | $t$-statistic |
|---|---|---|---|---|
| $\omega^{(1)}$ | $B_t$ | -1.857 | 0.116 | -15.959 |
| $\zeta^{(1)}$ | $B_t$ | 0.315 | 0.071 | 4.432 |
| $\eta^{(1)}$ | $B_t$ | 1.061 | 0.094 | 11.231 |
| $\delta^{(1)}$ | $B_t$ | -3.051 | 0.775 | -3.938 |
| $\phi_B^{(1)}$ | $B_t$ | 0.934 | 0.057 | 16.475 |
| $\alpha^{(1)}$ | $B_t$ | 0.292 | 0.086 | 3.390 |
| $\omega^{(1)}$ | $S_t$ | 2.007 | 0.043 | 46.979 |
| $\zeta^{(1)}$ | $S_t$ | 0.134 | 0.031 | 4.256 |
| $\eta^{(1)}$ | $S_t$ | 0.361 | 0.036 | 10.094 |
| $\delta^{(1)}$ | $S_t$ | -4.168 | 0.898 | -4.641 |
| $\phi_S^{(1)}$ | $S_t$ | 0.068 | 0.026 | 2.661 |
| $\alpha^{(1)}$ | $S_t$ | 0.108 | 0.014 | 7.847 |
| $\omega^{(2)}$ | $B_t$ | 1.581 | 0.037 | 43.041 |
| $\zeta^{(2)}$ | $B_t$ | -0.057 | 0.031 | -1.820 |
| $\eta^{(2)}$ | $B_t$ | 0.417 | 0.036 | 11.472 |
| $\delta^{(2)}$ | $B_t$ | -0.533 | 0.244 | -2.187 |
| $\phi_B^{(2)}$ | $B_t$ | 0.202 | 0.022 | 9.252 |
| $\alpha^{(2)}$ | $B_t$ | 0.320 | 0.022 | 14.410 |
| $\omega^{(2)}$ | $S_t$ | 1.246 | 0.053 | 23.535 |
| $\zeta^{(2)}$ | $S_t$ | -0.067 | 0.035 | -1.882 |
| $\eta^{(2)}$ | $S_t$ | 0.301 | 0.038 | 7.857 |
| $\delta^{(2)}$ | $S_t$ | -0.143 | 0.130 | -1.100 |
| $\phi_S^{(2)}$ | $S_t$ | 0.184 | 0.030 | 6.213 |
| $\alpha^{(2)}$ | $S_t$ | 0.390 | 0.031 | 12.665 |
| $\omega^{(3)}$ | $B_t$ | 0.701 | 0.052 | 13.545 |
| $\zeta^{(3)}$ | $B_t$ | 0.085 | 0.058 | 1.481 |
| $\eta^{(3)}$ | $B_t$ | 0.180 | 0.062 | 2.895 |
| $\delta^{(3)}$ | $B_t$ | 0.169 | 0.398 | 0.425 |
| $\phi_B^{(3)}$ | $B_t$ | 0.261 | 0.042 | 6.148 |
| $\alpha^{(3)}$ | $B_t$ | 0.652 | 0.071 | 9.147 |
| $\omega^{(3)}$ | $S_t$ | 0.843 | 0.053 | 15.750 |
| $\zeta^{(3)}$ | $S_t$ | 0.118 | 0.050 | 2.377 |
| $\eta^{(3)}$ | $S_t$ | 0.422 | 0.055 | 7.666 |
| $\delta^{(3)}$ | $S_t$ | 0.190 | 0.266 | 0.714 |
| $\phi_S^{(3)}$ | $S_t$ | 0.159 | 0.043 | 3.701 |
| $\alpha^{(3)}$ | $S_t$ | 0.457 | 0.057 | 8.051 |
| $p_{11}$ | - | 0.3453 | 0.019 | 8.553 |
| $p_{21}$ | - | 0.6543 | 0.019 | 10.770 |
| $p_{31}$ | - | 0.0004 | - | - |
| $p_{12}$ | - | 0.2551 | 0.011 | 6.652 |
| $p_{22}$ | - | 0.7335 | 0.011 | 25.567 |
| $p_{32}$ | - | 0.0114 | - | - |
| $p_{13}$ | - | 0.0112 | 0.002 | -78.076 |
| $p_{23}$ | - | 0.0027 | 0.001 | -419.492 |
| $p_{33}$ | - | 0.9861 | - | - |

| | | | | |
|---|---|---|---|---|
| $\ln \mathcal{L}_I(\theta)$ | -8297.1 | $\chi^2_{LB}$ | $B_t$ | 79.1 |
| $T$ | 1715 | | $S_t$ | 75.2 |
| $AIC$ | 16684.3 | $\chi^2_{CM}$ | $B_t$ | 39.1 |
| $BIC$ | 16929.4 | | $S_t$ | 40.1 |

$\chi^2_{LB}$ is computed for 50 lags. The corresponding tabulated 99% (95%) critical value for 50 degrees of freedom is 76.2 (67.5). $\chi^2_{CM}$ is computed for 12 cells, $(0,\dots,10, 11+)$. The corresponding tabulated 99% (95%) critical value for 11 degrees of freedom is 24.7 (19.7). The $t$-statistic for the regime probabilities corresponds to $H_0 : p_{ji} = \pi_j$.

**Table 5.12.** Parameter estimates KO

| Parameter | Equation | Estimate | Std.-error | $t$-statistic |
|---|---|---|---|---|
| $\omega^{(1)}$ | $B_t$ | -0.019 | 0.049 | -0.385 |
| $\zeta^{(1)}$ | $B_t$ | 0.215 | 0.040 | 5.384 |
| $\eta^{(1)}$ | $B_t$ | 0.204 | 0.039 | 5.200 |
| $\delta^{(1)}$ | $B_t$ | -11.285 | 0.898 | -12.568 |
| $\phi_B^{(1)}$ | $B_t$ | 0.478 | 0.033 | 14.584 |
| $\alpha^{(1)}$ | $B_t$ | 0.186 | 0.038 | 4.918 |
| $\omega^{(1)}$ | $S_t$ | 1.525 | 0.038 | 40.084 |
| $\zeta^{(1)}$ | $S_t$ | 0.069 | 0.031 | 2.243 |
| $\eta^{(1)}$ | $S_t$ | 0.187 | 0.035 | 5.325 |
| $\delta^{(1)}$ | $S_t$ | -2.855 | 0.677 | -4.217 |
| $\phi_S^{(1)}$ | $S_t$ | 0.116 | 0.028 | 4.154 |
| $\alpha^{(1)}$ | $S_t$ | 0.135 | 0.019 | 7.220 |
| $\omega^{(2)}$ | $B_t$ | 1.512 | 0.034 | 44.660 |
| $\zeta^{(2)}$ | $B_t$ | 0.048 | 0.034 | 1.404 |
| $\eta^{(2)}$ | $B_t$ | 0.181 | 0.037 | 4.883 |
| $\delta^{(2)}$ | $B_t$ | -1.046 | 0.213 | -4.908 |
| $\phi_B^{(2)}$ | $B_t$ | 0.181 | 0.024 | 7.573 |
| $\alpha^{(2)}$ | $B_t$ | 0.321 | 0.026 | 12.560 |
| $\omega^{(2)}$ | $S_t$ | 0.780 | 0.042 | 18.344 |
| $\zeta^{(2)}$ | $S_t$ | 0.161 | 0.040 | 3.996 |
| $\eta^{(2)}$ | $S_t$ | 0.241 | 0.037 | 6.476 |
| $\delta^{(2)}$ | $S_t$ | -0.090 | 0.190 | -0.476 |
| $\phi_S^{(2)}$ | $S_t$ | 0.216 | 0.032 | 6.701 |
| $\alpha^{(2)}$ | $S_t$ | 0.356 | 0.036 | 9.806 |
| $\omega^{(3)}$ | $B_t$ | 1.246 | 0.028 | 44.858 |
| $\zeta^{(3)}$ | $B_t$ | 0.055 | 0.026 | 2.091 |
| $\eta^{(3)}$ | $B_t$ | 0.223 | 0.026 | 8.553 |
| $\delta^{(3)}$ | $B_t$ | -0.805 | 0.175 | -4.608 |
| $\phi_B^{(3)}$ | $B_t$ | -0.024 | 0.023 | -1.075 |
| $\alpha^{(3)}$ | $B_t$ | 0.099 | 0.016 | 6.259 |
| $\omega^{(3)}$ | $S_t$ | 0.712 | 0.032 | 22.250 |
| $\zeta^{(3)}$ | $S_t$ | 0.085 | 0.030 | 2.846 |
| $\eta^{(3)}$ | $S_t$ | 0.354 | 0.032 | 11.144 |
| $\delta^{(3)}$ | $S_t$ | -0.434 | 0.147 | -2.955 |
| $\phi_S^{(3)}$ | $S_t$ | 0.124 | 0.027 | 4.654 |
| $\alpha^{(3)}$ | $S_t$ | 0.098 | 0.022 | 4.461 |
| $p_{11}$ | - | 0.4448 | 0.013 | 14.087 |
| $p_{21}$ | - | 0.3695 | 0.010 | 1.181 |
| $p_{31}$ | - | 0.1857 | - | - |
| $p_{12}$ | - | 0.2683 | 0.009 | 1.047 |
| $p_{22}$ | - | 0.7103 | 0.009 | 39.923 |
| $p_{32}$ | - | 0.0213 | - | - |
| $p_{13}$ | - | 0.1252 | 0.005 | -24.932 |
| $p_{23}$ | - | 0.0202 | 0.001 | -410.673 |
| $p_{33}$ | - | 0.8546 | - | - |

| | | | | |
|---|---|---|---|---|
| $\ln \mathcal{L}_I(\theta)$ | -7693.6 | $\chi^2_{LB}$ | $B_t$ | 42.8 |
| $T$ | 1715 | | $S_t$ | 51.7 |
| $AIC$ | 15477.3 | $\chi^2_{CM}$ | $B_t$ | 19.1 |
| $BIC$ | 15722.4 | | $S_t$ | 12.3 |

$\chi^2_{LB}$ is computed for 50 lags. The corresponding tabulated 99% (95%) critical value for 50 degrees of freedom is 76.2 (67.5). $\chi^2_{CM}$ is computed for 12 cells, (0,...,
10, 11+). The corresponding tabulated 99% (95%) critical value for 11 degrees of freedom is 24.7 (19.7). The $t$-
statistic for the regime probabilities corresponds to $H_0$ :
$p_{ji} = \pi_j$.

**Table 5.13.** Parameter estimates XON

| Parameter | Equation | Estimate | Std.-error | $t$-statistic |
|---|---|---|---|---|
| $\omega^{(1)}$ | $B_t$ | -1.827 | 0.096 | -18.955 |
| $\zeta^{(1)}$ | $B_t$ | 0.259 | 0.097 | 2.660 |
| $\eta^{(1)}$ | $B_t$ | 1.300 | 0.115 | 11.337 |
| $\delta^{(1)}$ | $B_t$ | 0.213 | 0.187 | 1.139 |
| $\phi_B^{(1)}$ | $B_t$ | 0.719 | 0.062 | 11.613 |
| $\alpha^{(1)}$ | $B_t$ | 0.109 | 0.099 | 1.110 |
| $\omega^{(1)}$ | $S_t$ | 1.232 | 0.036 | 34.005 |
| $\zeta^{(1)}$ | $S_t$ | 0.250 | 0.045 | 5.602 |
| $\eta^{(1)}$ | $S_t$ | 0.307 | 0.041 | 7.517 |
| $\delta^{(1)}$ | $S_t$ | -0.760 | 0.200 | -3.798 |
| $\phi_S^{(1)}$ | $S_t$ | -0.078 | 0.033 | -2.333 |
| $\alpha^{(1)}$ | $S_t$ | 0.051 | 0.024 | 2.151 |
| $\omega^{(2)}$ | $B_t$ | 1.043 | 0.033 | 31.678 |
| $\zeta^{(2)}$ | $B_t$ | 0.179 | 0.035 | 5.117 |
| $\eta^{(2)}$ | $B_t$ | 0.357 | 0.037 | 9.673 |
| $\delta^{(2)}$ | $B_t$ | -0.707 | 0.298 | -2.373 |
| $\phi_B^{(2)}$ | $B_t$ | 0.157 | 0.025 | 6.270 |
| $\alpha^{(2)}$ | $B_t$ | 0.296 | 0.028 | 10.626 |
| $\omega^{(2)}$ | $S_t$ | 0.247 | 0.038 | 6.454 |
| $\zeta^{(2)}$ | $S_t$ | 0.094 | 0.044 | 2.144 |
| $\eta^{(2)}$ | $S_t$ | 0.207 | 0.045 | 4.559 |
| $\delta^{(2)}$ | $S_t$ | -0.843 | 0.357 | -2.362 |
| $\phi_S^{(2)}$ | $S_t$ | 0.425 | 0.036 | 11.780 |
| $\alpha^{(2)}$ | $S_t$ | 0.411 | 0.049 | 8.426 |
| $\omega^{(3)}$ | $B_t$ | 1.192 | 0.043 | 27.512 |
| $\zeta^{(3)}$ | $B_t$ | -0.038 | 0.041 | -0.912 |
| $\eta^{(3)}$ | $B_t$ | 0.228 | 0.041 | 5.564 |
| $\delta^{(3)}$ | $B_t$ | -13.145 | 1.904 | -6.904 |
| $\phi_B^{(3)}$ | $B_t$ | 0.123 | 0.030 | 4.070 |
| $\alpha^{(3)}$ | $B_t$ | 0.533 | 0.043 | 12.330 |
| $\omega^{(3)}$ | $S_t$ | 0.945 | 0.042 | 22.749 |
| $\zeta^{(3)}$ | $S_t$ | -0.089 | 0.047 | -1.884 |
| $\eta^{(3)}$ | $S_t$ | 0.240 | 0.048 | 4.997 |
| $\delta^{(3)}$ | $S_t$ | -12.816 | 0.388 | -33.072 |
| $\phi_S^{(3)}$ | $S_t$ | 0.140 | 0.035 | 3.941 |
| $\alpha^{(3)}$ | $S_t$ | 0.647 | 0.052 | 12.321 |
| $p_{11}$ | - | 0.5592 | 0.018 | 21.319 |
| $p_{21}$ | - | 0.4100 | 0.018 | 0.433 |
| $p_{31}$ | - | 0.0308 | - | - |
| $p_{12}$ | - | 0.1827 | 0.009 | 1.769 |
| $p_{22}$ | - | 0.8006 | 0.009 | 44.254 |
| $p_{32}$ | - | 0.0167 | - | - |
| $p_{13}$ | - | 0.0000 | 0.000 | -46971.814 |
| $p_{23}$ | - | 0.0275 | 0.002 | -248.254 |
| $p_{33}$ | - | 0.9725 | - | - |

| | | | | |
|---|---|---|---|---|
| $\ln \mathcal{L}_I(\theta)$ | -7131.5 | $\chi^2_{LB}$ | $B_t$ | 76.9 |
| $T$ | 1715 | | $S_t$ | 60.8 |
| $AIC$ | 14352.9 | $\chi^2_{CM}$ | $B_t$ | 32.4 |
| $BIC$ | 14598.1 | | $S_t$ | 26.5 |

$\chi^2_{LB}$ is computed for 50 lags. The corresponding tabulated 99% (95%) critical value for 50 degrees of freedom is 76.2 (67.5). $\chi^2_{CM}$ is computed for 12 cells, (0,..., 10, 11+). The corresponding tabulated 99% (95%) critical value for 11 degrees of freedom is 24.7 (19.7). The $t$-statistic for the regime probabilities corresponds to $H_0$: $p_{ji} = \pi_j$.

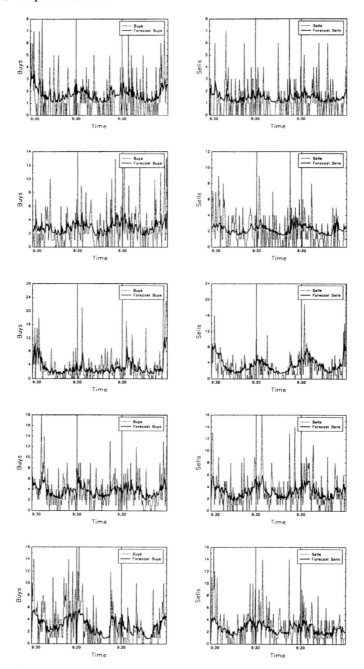

**Fig. 5.4.** Unconditional forecasts on three consecutive trading days (August 19 - August 21, 1996). Left panel: Buys, right panel: Sells. From top to bottom: Boeing, Disney, IBM, Coca-Cola, Exxon.

### 5.4.3 Specification Tests

Turning to the results of our specification tests, we find, that the models selected by the $BIC$ tend to produce residuals, that are serially uncorrelated at least at the 1% significance level for both time series. There are only two exceptions, the Pearson residuals for IBM and XON buys, for which the Ljung-Box statistics computed for 50 lags are very close to the 1% critical values, and furthermore the plots of the autocorrelation functions of both residual time series do not exhibit any specific serial dependence patterns, except for occasional crossings of the bounds of the associated 95%-confidence intervals that appear to be purely coincidental in nature, see Fig. 5.5. In line with this impression, the Ljung-Box statistics for lag orders up to 30 and for 100 lags do not indicate rejection of the Null hypothesis of no residual autocorrelation in both cases.

With respect to our distributional assumptions, we find the evidence in favor of the selected model specifications to be less convincing however. We conducted separate conditional moment tests for each of the two dependent variables for the 12 cells defined by $(y = 0, y = 1, \ldots, y = 10, y \geq 11)$, and find that the Null hypothesis of correct specification of the density is not rejected at conventional significance levels only in five cases, the time series of sells for BA, DIS, and KO, as well as for KO buys. In all other cases, the models we estimated do not seem to describe the observed densities of the dependent variables very well. Occasionally, one of the model specifications that was rejected by the $BIC$ would improve on the p-value for one or the other variable, but we did not find any model specification that does better for both dependent variables at the same time.

Despite of this apparent lack of fit with regard to the sampling densities, we stick to the model specifications selected by the use of the information criteria, since we based our subsequent statistical inferences on the QML-estimator of the information matrix proposed by White (1982), which is robust against misspecification of the sampling density and can therefore be used to compute asymptotically unbiased estimates of the standard errors of individual parameter estimates. It should be kept in mind though, that the goodness of fit of our estimated models might be improved with respect to the underlying distributional assumptions, though the mere fact of rejection by the conditional moment test does not indicate in which direction to proceed. Figure 5.6 contains plots of the residual function along with the corresponding 95% confidence intervals. The plots of the residual functions for the selected models appear do be virtually flat, and the corresponding confidence intervals all encompass a horizontal line through zero, except for DIS buys and IBM sells, for which the confidence intervals do not encompass the horizontal line at the origin $(y = 0)$ only. Note, that the confidence intervals widen and the plot of the residual functions generally becomes wiggly as the values of the dependent variable become larger. As noted in Sect. 4.2.2, this is to a large extent a result of the sparsity of sample observations that fall into this region of the variable's support. Although the fit of our models to the data appears to be better for small realizations of the trading events the residual function fluctuates around zero even in sparse regions of the support thus supporting the impression, that the chosen 3-regime models based on the Negbin 2 regression model perform quite well.

Figure 5.7 presents plots of the empirical density of the trading events versus the expected density implied by the negative binomial 3-regime Markov switching models we estimated. Note, that the fit of the estimated density in general matches the shapes of the empirical density for both variables quite well. Nevertheless, there are also apparent deviations between the two densities and these tend to be quite large in some cases (DIS buys, IBM sells). In general the distance between the expected and the empirical density is larger for small realizations of the dependent variables, say in the range $0 \leq y \leq 10$. Noting, that the plot of the residual function reproduced in Fig. 5.6 contains essentially the same information as the density plot[25] in Fig. 5.7, this seems to contradict our former impression, that the fit of our models is better for small values of the dependent variables. This seemingly contradictory implication is a result of the different implicit measures of the distance between the two densities. First note, that the residual function may be rewritten as follows

$$c(y; \hat{\theta}) = [\tilde{f}(y) - f(y \mid \hat{\theta})] \cdot \frac{1}{f(y \mid \hat{\theta})}.$$

The visual impression from Fig. 5.7, as well as the conditional moment test statistic, are based on measures of the *absolute* distance of the two densities, i.e. $\mid \tilde{f}(y) - f(y \mid \hat{\theta}) \mid$, with simultaneous consideration of the sampling variation of the implied expected density in the case of the conditional moment test, thereby accentuating discrepancies between the two densities that are either large in absolute value or are affected by a large sampling variation of the estimated parameters of the distribution as measured by the score function $s_t$ that enters the conditional moment test statistic. The residual function on the other hand measures the *relative* distance between the two densities, thus implicitly downweighting discrepancies between the densities in regions of the support where the expected density of observing a particular realization $f(y \mid \hat{\theta})$ is high and accentuating discrepancies between the densities when the expected density is low.

In our case, the absolute distance between the two densities appears to be large for smaller values of the dependent variables, while at the same time these small values have a high expected probability of being observed, so the effect of large deviations between the two densities in the range $0 \leq y \leq 10$ is being offset by the low weights these deviations receive when they enter the

---

[25] Recall that the residual function is defined as the ratio of the empirical density $\tilde{f}(y)$ to the expected density implied by the estimated model $f(y \mid \hat{\theta})$ minus one, see Sect. 4.2.2.

residual function $c(y; \hat{\theta})$. Deb and Trivedi (2002) report that the conditional moment test tends to overreject the true null hypothesis at conventional significance levels. Therefore, they suggest to use conditional moment tests only to rank different model specifications by associated p-values, so that the model associated with the largest p-value would be preferred.

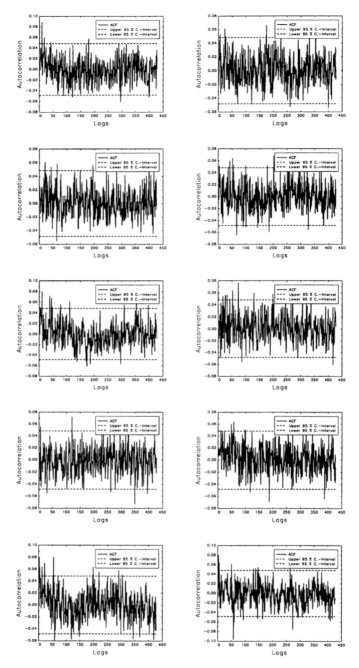

**Fig. 5.5.** ACF of the Pearson residuals. Left panel: Buys, right panel: Sells. From top to bottom: Boeing, Disney, IBM, Coca-Cola, Exxon.

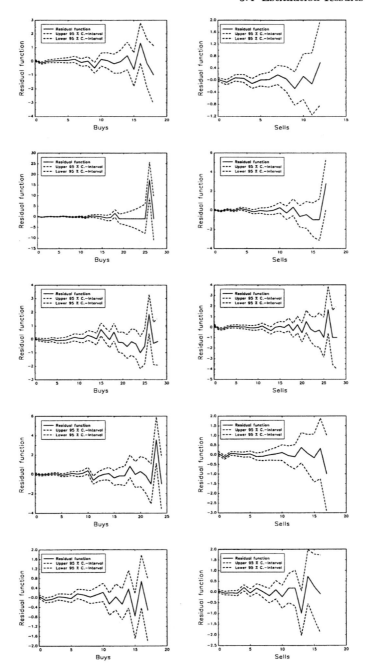

**Fig. 5.6.** Residual function. Left panel: Buys, right panel: Sells. From top to bottom: Boeing, Disney, IBM, Coca-Cola, Exxon.

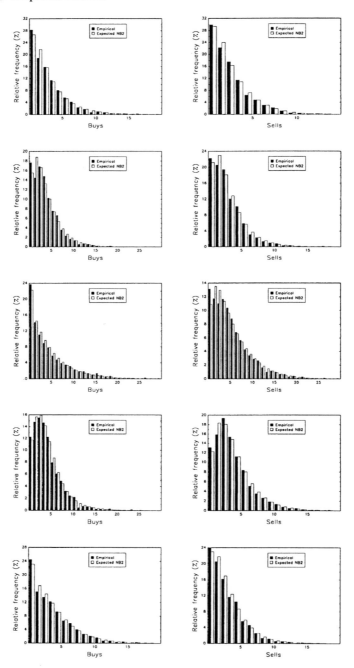

**Fig. 5.7.** Empirical and estimated density. Left panel: Buys, right panel: Sells. From top to bottom: Boeing, Disney, IBM, Coca-Cola, Exxon.

### 5.4.4 Classification of Regimes

In order to interpret our models in terms of the underlying economic model we examine the regime specific properties of these models more closely in this Section. In Table 5.14 we present the sample means of the regime specific forecasts $\hat{y}_{tjk}$ and conditional variances $\hat{v}_{tjk}$ along with the eigenvalues of the estimated transition matrix $P$, the expected regime durations and the stationary probabilities of the associated Markov chains. Note, that it would be hard to derive an explicit expression of the unconditional expectation of the dependent variable under regime $j$, denoted by $E(y_{tjk} \mid r = j; \theta)$, as a function of estimated parameters, because of the inclusion of exogenous explanatory variables in the regression function (5.1). Nevertheless, we can make good use of the following relationship between conditional and unconditional expectations[26]

$$E(y_{tjk} \mid r = j; \theta) = E[E(y_{tjk} \mid r = j, \mathcal{F}_t; \theta)]. \tag{5.2}$$

Replacing population moments by sample moments on the right hand side of (5.2) yields

$$\hat{y}_{jk} = \frac{1}{T} \cdot \sum_{t=1}^{T} \hat{y}_{tjk}, \tag{5.3}$$

where $\hat{y}_{jk}$ denotes the implied estimate of $E(y_{tjk} \mid r = j; \theta)$.

Comparing the sample means of the regime specific forecasts $\hat{y}_{tjk}$, reproduced in Table 5.14 for all stocks in our sample, thus allows us to classify the 3 regimes. Recall, that the sequential trade framework implies, that under the bad news regime we expect to observe more sells than buys, under good news we expect more buys than sells, and under no news we would expect to have approximately as many buys as sells. The sample averages of the sell forecasts under regime 1 are all higher than the corresponding figures for the buys, implying that regime 1 is the bad news regime. Under regime 2 the sample averages of the buy forecasts are higher than for the sells, indicating that regime 2 represents good news. Under regime 3 the averages appear to be roughly equal, or to be more precise, the absolute difference between the average forecasts is smallest among all three regimes, with the average number of buys slightly higher than the average number of sells for all stocks except IBM for which the average number of sells is slightly higher than for buys. This suggests that regime 3 describes the behavior of the market in the absence of information events, and thus represents the no news regime.

Comparing the corresponding sample averages of the conditional variances, we find, that the variance of the sells appears to be higher under bad news, while the opposite is true for the good news regime and the no news regime. This salient feature holds for all stocks in our sample, again with the exception of IBM, where the variance of the sells is higher than the corresponding

---

[26] See Spanos (1986), p. 126.

**Table 5.14.** Properties of the regimes

| Statistic | Variable | Regime 1 | Regime 2 | Regime 3 |
|-----------|----------|----------|----------|----------|
| | | Boeing | | |
| Mean | $B_t$ | 0.82 | 3.83 | 1.78 |
| | $S_t$ | 4.09 | 1.58 | 1.44 |
| Variance | $B_t$ | 1.08 | 4.93 | 3.23 |
| | $S_t$ | 13.71 | 3.07 | 2.50 |
| $EV$ | - | 0.00 | 0.96 | 1.00 |
| $E(D_j)$ | - | 1.39 | 3.31 | 39.49 |
| $\pi_j$ | - | 0.17 | 0.39 | 0.44 |
| | | Disney | | |
| Mean | $B_t$ | 1.57 | 5.07 | 2.76 |
| | $S_t$ | 4.40 | 1.85 | 2.06 |
| Variance | $B_t$ | 2.31 | 8.05 | 4.64 |
| | $S_t$ | 10.87 | 3.61 | 3.74 |
| $EV$ | - | 0.16 | 1.00 | 0.97 |
| $E(D_j)$ | - | 1.89 | 3.06 | 72.67 |
| $\pi_j$ | - | 0.16 | 0.28 | 0.56 |
| | | IBM | | |
| Mean | $B_t$ | 0.99 | 6.32 | 2.72 |
| | $S_t$ | 8.30 | 4.51 | 3.03 |
| Variance | $B_t$ | 1.97 | 11.46 | 5.44 |
| | $S_t$ | 37.83 | 19.12 | 7.87 |
| $EV$ | - | 0.09 | 1.00 | 0.98 |
| $E(D_j)$ | - | 1.53 | 3.75 | 71.91 |
| $\pi_j$ | - | 0.18 | 0.45 | 0.37 |
| | | Coca-Cola | | |
| Mean | $B_t$ | 1.73 | 5.49 | 3.42 |
| | $S_t$ | 5.10 | 2.75 | 2.37 |
| Variance | $B_t$ | 2.44 | 9.79 | 7.28 |
| | $S_t$ | 14.79 | 3.57 | 2.96 |
| $EV$ | - | 0.21 | 1.00 | 0.80 |
| $E(D_j)$ | - | 1.80 | 3.45 | 6.88 |
| $\pi_j$ | - | 0.26 | 0.36 | 0.38 |
| | | Exxon | | |
| Mean | $B_t$ | 0.57 | 3.34 | 3.62 |
| | $S_t$ | 3.40 | 1.77 | 2.82 |
| Variance | $B_t$ | 0.66 | 3.98 | 7.75 |
| | $S_t$ | 8.52 | 3.79 | 8.29 |
| $EV$ | - | 0.38 | 1.00 | 0.95 |
| $E(D_j)$ | - | 2.27 | 5.02 | 36.37 |
| $\pi_j$ | - | 0.17 | 0.40 | 0.43 |

$EV$ = eigenvalue of the transition matrix, $E(D_j)$ = expected duration of regime $j$, $\pi_j$ = ergodic probability of regime $j$.

variance of the buys under all 3 regimes, and XON, where the variance of the sells is higher than the variance of the buys under the no news regime. The eigenvalues of the transition matrix indicate that the implied Markov chains for the regime variables are indeed ergodic, as we always find one eigenvalue to be equal to one, while the remaining two are positive and smaller than one in absolute value, and thus lie inside the unit circle.[27]

With respect to the expected duration of the regimes, our estimates indicate that in all cases the no news regime has the longest expected duration. There are however significant differences in the magnitude of the durations between the stocks in our sample. For two stocks (IBM, DIS) we find that the expected duration of a no news regime is close to one trading day, which is equal to 78 trading intervals of 5 minutes length. The expected duration of the no news regime for BA and XON is roughly equal to half a trading day (39 trading intervals), while the no news regime for KO appears to last only roughly 35 minutes (7 trading intervals) on average. Furthermore, we observe that for all stocks the expected duration of the bad news regime is the shortest, lasting less than 2 trading intervals or 10 minutes on average, while the good news regime has a slightly higher expected duration of roughly 3.5 trading intervals (or about 17.5 minutes). Only XON appears to have slightly more persistent good and bad news regimes, with average durations of 2.27 trading intervals (11.35 minutes) for the bad news regime and 5.02 trading intervals (25.1 minutes) under good news. This feature appears to be consistent with our earlier finding, that the number of buys is larger than the number of sells for the stocks in our sample, compare Tables 5.2 and 5.3.

Another noteworthy result implied by our estimates, is that the ergodic regime probabilities $\pi_j$ tend to be highest for the no news regime for all our stocks, with shares between 37% (IBM) and 56% (DIS). An exception is the IBM stock, for which we find a higher unconditional probability of being in the good news regime, than in the no news regime. This may be interpreted as evidence that a large fraction of all trades represents transaction demand by uninformed traders. However, with the exception of the DIS stock, the unconditional probabilities of being in any of the two news regimes is higher than 50%, thus indicating, that information was present in the market during more than half of our sample period. Furthermore we find, that when an information event has occurred, it was more likely to be good news, than bad news, as the ergodic probabilities for good news are higher than the corresponding probabilities for bad news for all stocks in our sample.

Figure 5.8 illustrates this feature by showing the evolution of the smoothed regime probabilities $\hat{\xi}_{t|T}^{(j)}$ during three consecutive trading days (August 19 until August 21, 1996). Note, that the plot shows the cumulative regime probabilities, i.e. the probability of being in the bad news regime 1 is indicated by the dark grey area, the probability of being in the good news regime 2 is indicated by the lighter grey area, and the probability of being in the no news regime 3 is represented by the white area in Fig. 5.8. The graphs reveal that the informational epochs during these three days were quite different for the stocks in our sample. While the trading process for the BA stock is basically characterized by the absence of information events, with only two minor

---

[27] Compare again Appendix A.8.

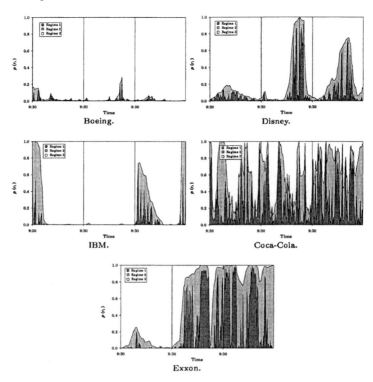

**Fig. 5.8.** Smoothed regime probabilities on three consecutive trading days (August 19 - August 21, 1996).

episodes at the beginning of the first and during the afternoon of the second trading day, both DIS and IBM appear to be going through three major informational epochs with differing length. The IBM stock has informational epochs at the beginning of the first and third trading day and near the close of the third day, that appear to be essentially episodes of good news.

In the case of DIS, there is one epoch with a peak around lunchtime of the first day, during which the regime seems to be predominantly in the good news regime, with frequent but short interruptions of switching into bad news. Note however, that the cumulative probability of being in either of the two news regimes is never any higher than 20%, thus casting some doubts on the significance of the informational content of the trading process in this period. The second epoch begins in the afternoon of the second day and ends approximately one hour before the close. This epoch as well as the third one, which begins around noon and ends again approximately one hour before the close, appears to be more significant, given the large likelihood of not being in the no news regime. Again, these episodes can be characterized by the presence of good news, with several large, but short interruptions during which the bad news state is being entered.

This picture of frequent switching between good and bad news is also evident in the graphs of the two remaining stocks, most prominently in the case of KO, where the information regime appears to be switching frequently between all three regimes in the course of these there days. The KO stock goes through a very turbulent phase, in which the market perception appears to be driven by an insecurity about the future prospects of this share. In the case of XON, we observe a longer informational epoch during the second and third day, again associated with frequent switches between good and bad news, while the first trading day is characterized by the absence of news, with only a minor good news episode around noon.

In Fig. 5.9 we additionally present plots of the regime specific forecasts of the trading process for our sample of stocks during the same period of time. Note, that the regime specific forecasts are predominantly driven by the U-shaped seasonal pattern, and the associated trading momentum that was modelled by the autoregressive terms in each regression equation. It is apparent, that the forecasts for regime 1, the bad news regime, and regime two, the good news regime are well separated from each other, in the sense that they operate on different levels for both regimes and equations. For the time series of buys, we observe that the level of the forecasts is generally high under regime 2 (good news) and low under regime 1 (bad news). The opposite relation between the two regime specific forecasts holds true for the time series of sells. While this pattern is more or less the same for all stocks in our sample, the forecasts under the third, no news regime behave differently among stocks. In some cases (BA, IBM, and KO buys) the level of the forecasts lies between the levels of the good resp. bad news regimes, in other cases it seems to follow more closely the level of the good news regime (XON buys, BA, DIS, and KO sells) or the bad news regime (DIS buys). In one case (XON sells) the plots of the forecasts of all three regimes intersect with each other several times. Such crossings of the regime specific levels of trading are not possible within the original EKOP-framework, which would predict that forecasts of the trading activity under no news are strictly lower than under any of the two other regimes.[28]

The occurrence of these frequent switches between the good and bad news regime, and the level crossings of the associated trade arrival rates may be interpreted in a number of ways. Previous empirical research that was based on the basic sequential trade framework does not predict such short-lived episodes explicitly. Recall that most of the empirical work was based on the assumption that each trading day should be governed either by the absence of news or by the occurrence of a singular information event, that drives the

---

[28] To be more precise, the EKOP-model predicts, that the number of sells under bad news is strictly higher than the number of sells under no news, and the number of buys under good news is strictly higher than the number of buys under no news.

market in one direction or the other. There is even some evidence that the trading process is correlated on a daily basis, implying that informational epochs might as well have average durations of several days or longer.[29] In the light of this interpretation the frequent switching activity between state 1 and 2 might be caused by a weakness of our statistical model that has difficulties to identify the information regime unambiguously. However, we think that the economic model should not be taken literally at this particular point, since there is nothing in the sequential trade model that would prevent us from taking the period of trading to be shorter than one complete trading day either.

An alternative view is to relate these phases of frequent switching between the bad and good news regimes to the interaction between the market maker and speculative traders. From this point of view, the seemingly arbitrariness of switching might be caused by the reaction of the market maker to a sequence of one sided trades, that he suspects of being information based or that may cause him troubles to uphold a sufficient inventory of stocks or cash. In such cases, the sequential trade framework predicts, that the market maker will widen the spread, e.g. by lowering (increasing) the bid (ask) price, when he suspects that bad (good) news are present. Given the apparent heterogeneity of the trading population, this might in turn encourage some traders to buy against the trend, which in some cases may lead to *overshooting* market reactions, in the sense that an initially higher number of sells than buys in one period of time is followed by a trade imbalance in the opposite direction, i.e. more buys than sells in the next interval. It is the very core of the sequential trade framework, that the presence and nature of any information reveals itself in time through the occurrence of such trade imbalances.

Taken together, our evidence describes a trading process, that operates in absence of information events for relatively long periods of time, driven mainly by uninformed transaction demands, with frequent, but relatively short interruptions caused by the occurrence of information events, during which a higher level of trading activity on one or even both sides of the market may be observed. These outbursts of activity tend to be driven by additional transaction demand of informed traders. Typically we observe overshooting market reactions during informational epochs, i.e. they are characterized by frequent switches between the good and bad news regime rather than by a persistent stay in either one of the two information regimes. Furthermore, we find that during these informational epochs not only the level but also the variability of trading activity changes systematically. Such informational epochs seem to differ from normal trading in a number of aspects apart from the pure level of activity.

---

[29] Typically, such findings are attributed to long run time trends in the arrival rates of informed and uninformed traders, not to serial dependence in the information process, see e.g. Lei and Wu (2001), Easley, Hvidkjaer, and O'Hara (2001), and Easley, Engle, O'Hara, and Wu (2002).

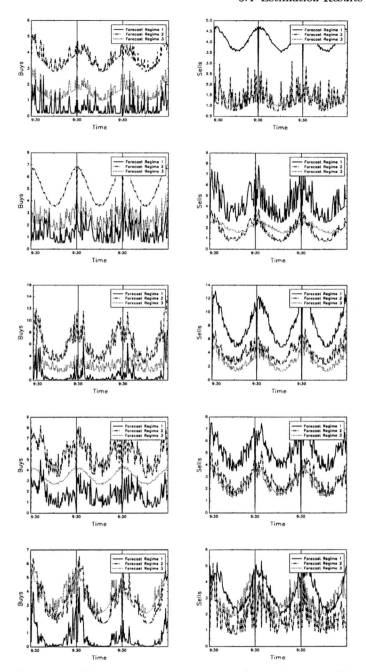

**Fig. 5.9.** Regime specific forecasts on three consecutive trading days (August 19 - August 21, 1996). Left panel: Buys, right panel: Sells. From top to bottom: Boeing, Disney, IBM, Coca-Cola, Exxon.

### 5.4.5 Testing Parameter Restrictions

In order to compare our preferred model specifications with the original EKOP-model, we will report the results of some test statistics, that validate restrictions among our parameter estimates. In order to conduct these tests we employ the Wald test, based on the QML-estimator of the information matrix.[30] As laid out in Sect. 4.1.2, a reduced form econometric model variant of the original EKOP-model is nested as a special case in our model specifications. This variant can be obtained by imposing the restriction, that the sell arrival rate under bad news and the buy arrival rate under good news are the same, representing the joint arrival rates of informed and uninformed traders in the presence of information events, and when additionally the buy arrival rate under bad news, the sell arrival rate under good news and both the buy and the sell arrival rates under no news, representing the marginal arrival rates of uninformed traders only, are restricted to be the same.

The validity of these equalities can in principle be tested by imposing equality restrictions among the corresponding sets of regime and equation specific regression parameter estimates. If these parameter restrictions hold true in our sample, the arrival rates of informed traders and uninformed traders across regimes would be as predicted by the original EKOP-model, despite the fact that we work with a Markov switching model instead of a static mixture model as EKOP did. However, in our model selection step we found, that an autoregressive specification of the regression function proved to yield the best fit in our sample. For this reason, the equality of e.g. the buy and sell arrival rates under the no news regime cannot hold in our model specification, even if the parameter estimates were exactly the same. This is a consequence of the fact that we let different variables enter the conditional mean specification in the two equations of our econometric model. The regression function for the sells depends on lagged sells, while the conditional mean of the buys depends on lagged buys only. Therefore our estimated models are a priori not conformable with a symmetric sequential trade model, in which buy and sell arrival rates of different types of traders are equal.

Instead, we can conduct tests on whether e.g. the seasonal pattern of the buy and sell arrival rates are the same when informed traders are present, whether the buy arrival rates of uninformed traders are the same under bad and no news, and whether the sell arrival rates of uninformed traders are the same under good and no news. If these restrictions would hold in our sample our econometric model would be conformable with the reduced form version of an asymmetric sequential trade model variant, similar in spirit to the model variants of Easley, Kiefer, and O'Hara (1996), Brown, Thomson, and Walsh (1999), and Easley, O'Hara, and Saar (2001), in which the arrival rates of uninformed buyers and sellers are not restricted to be the same.

We therefore conduct a total of 5 different parameter restriction tests, that are designed to investigate these issues. The corresponding Null hypothesis

---

[30] See Sect. 4.2.4 for details.

**Table 5.15.** Results of parameter restriction tests

| Firm | $g$ | $\alpha$ unrestricted | | $g$ | $\alpha$ restricted | |
|------|-----|------|------|-----|------|------|
| | | $\chi^2_W$ | p-value | | $\chi^2_W$ | p-value |
| | | | $H_1$ | | | |
| BA | 4 | 9.41 | 0.0516 | 5 | 121.39 | 0.0000 |
| DIS | 4 | 20.18 | 0.0005 | 5 | 22.04 | 0.0005 |
| IBM | 4 | 103.87 | 0.0000 | 5 | 221.45 | 0.0000 |
| KO | 4 | 6.62 | 0.1577 | 5 | 51.14 | 0.0000 |
| XON | 4 | 19.98 | 0.0005 | 5 | 90.89 | 0.0000 |
| | | | $H_2$ | | | |
| BA | 5 | 879.52 | 0.0000 | 6 | 907.29 | 0.0000 |
| DIS | 5 | 204.95 | 0.0000 | 6 | 248.31 | 0.0000 |
| IBM | 5 | 507.77 | 0.0000 | 6 | 511.60 | 0.0000 |
| KO | 5 | 787.58 | 0.0000 | 6 | 823.38 | 0.0000 |
| XON | 5 | 956.07 | 0.0000 | 6 | 1088.51 | 0.0000 |
| | | | $H_3$ | | | |
| BA | 5 | 26.03 | 0.0001 | 6 | 27.00 | 0.0001 |
| DIS | 5 | 98.65 | 0.0000 | 6 | 101.23 | 0.0000 |
| IBM | 5 | 101.90 | 0.0000 | 6 | 102.12 | 0.0000 |
| KO | 5 | 47.13 | 0.0000 | 6 | 94.27 | 0.0000 |
| XON | 5 | 700.08 | 0.0000 | 6 | 724.14 | 0.0000 |
| | | | $H_4$ | | | |
| BA | 10 | 967.34 | 0.0000 | 12 | 996.05 | 0.0000 |
| DIS | 10 | 302.32 | 0.0000 | 12 | 347.17 | 0.0000 |
| IBM | 10 | 633.07 | 0.0000 | 12 | 641.41 | 0.0000 |
| KO | 10 | 890.01 | 0.0000 | 12 | 1006.06 | 0.0000 |
| XON | 10 | 1604.17 | 0.0000 | 12 | 1750.25 | 0.0000 |
| | | | $H_5$ | | | |
| BA | 14 | 998.05 | 0.0000 | 17 | 1095.58 | 0.0000 |
| DIS | 14 | 319.85 | 0.0000 | 17 | 364.83 | 0.0000 |
| IBM | 14 | 794.55 | 0.0000 | 17 | 982.28 | 0.0000 |
| KO | 14 | 906.97 | 0.0000 | 17 | 1043.19 | 0.0000 |
| XON | 14 | 1821.38 | 0.0000 | 17 | 1991.82 | 0.0000 |

$g$ = number of restrictions.

are denoted as $H_1$ through $H_5$. We will employ two different variants of these tests. In the first variant the Null hypothesis are defined as follows

$$H_1 : \omega_S^{(1)} = \omega_B^{(2)} \cap \zeta_S^{(1)} = \zeta_B^{(2)} \cap \eta_S^{(1)} = \eta_B^{(2)} \cap \delta_S^{(1)} = \delta_B^{(2)}$$
$$H_2 : \omega_B^{(1)} = \omega_B^{(3)} \cap \zeta_B^{(1)} = \zeta_B^{(3)} \cap \eta_B^{(1)} = \eta_B^{(3)} \cap \delta_B^{(1)} = \delta_B^{(3)} \cap \phi_B^{(1)} = \phi_B^{(3)}$$
$$H_3 : \omega_S^{(2)} = \omega_S^{(3)} \cap \zeta_S^{(2)} = \zeta_S^{(3)} \cap \eta_S^{(2)} = \eta_S^{(3)} \cap \delta_S^{(2)} = \delta_S^{(3)} \cap \phi_S^{(2)} = \phi_S^{(3)}$$
$$H_4 : H_2 \cap H_3$$
$$H_5 : H_1 \cap H_2 \cap H_3,$$

i.e. we will only test equality restrictions with respect to the regression parameter estimates. Note, that we introduced indices in order to distinguish parameters that appear in the regression functions for buys from those that appear in the regression function for sells. Thus, for example $\omega_B^{(1)}$ denotes the constant term in the regression function for the buys under regime 1, while $\omega_S^{(1)}$ is the corresponding parameter appearing in the regression function for the sells under regime 1. The first hypothesis $H_1$ is a test on whether the seasonal components of the joint arrival rate of informed and uninformed traders are equal. Hypothesis $H_2$ is a test on whether the buy arrival rates of the uninformed traders are the same under the bad and no news regime, and $H_3$ is the corresponding test on whether the sell arrival rates of uninformed traders are equal under the good and the no news regime. The hypothesis $H_4$ is a joint test of equality of buy and sell arrival rates of the uninformed traders, i.e. whether $H_2$ and $H_3$ hold simultaneously, and $H_5$ additionally imposes equality of the seasonal components of the joint arrival rates of informed and uninformed traders, i.e. whether $H_1$ and $H_4$ hold at the same time.

In our second variant we additionally impose equality constraints between the corresponding dispersion parameter estimates $\alpha$. Thus, the $\alpha$-restricted variant of the equality test for the seasonal components of the joint arrival rates of informed and uninformed traders is $H_1 \cap \alpha_S^{(1)} = \alpha_B^{(2)}$, with analogous extensions for the other four tests. The basic sequential trade framework does not make any explicit prediction about these parameters since it assumes that all arrival rates follow a Poisson process, but given our discussion in Sect. 4.1.4 where we motivated the negative binomial regression model as a device to account for unobserved within-group heterogeneity, it is of interest to test whether the composition of the informed and uninformed trading population is the same across regimes.

In Table 5.15 we reproduce the results of these parameter tests. The columns entitled '$\alpha$ unrestricted' contain the test statistics $\chi_W^2$ and the corresponding p-values for the test concerning the regression parameters only, while the columns titled '$\alpha$ restricted' contain the results for the associated tests, when the corresponding regime- and equation-specific estimates of $\alpha$ are being restricted in addition. With the exception of the first hypothesis test for the BA and KO stocks in the $\alpha$ unrestricted version, that tests whether the seasonal pattern of the joint arrival rates of informed and uninformed traders under good and bad news are the same, all hypothesis tests are being rejected at conventional significance levels. These results reconfirm our earlier impression, that the original EKOP-model is to restrictive as a statistical model for intradaily time series of trade event counts.

# 6

# Conclusions

In this study we proposed a new framework for the estimation of sequential trade models. Our intention was to extend the scope of these models to the analysis of high frequency data sets based on intradaily sampling schemes. Most previous studies employed a daily sampling scheme thus neglecting a significant portion of the available information. We introduced several extensions of the basic sequential trade model for this purpose and laid out in detail the close connection between the inherent economic theory and the econometric methods we used. Furthermore, we showed how statistical tests can be used to test certain aspects of the underlying theory, that have not been assessed in previous empirical studies of sequential trade models.

We also introduced several new regression models for multivariate time series of count data based on mixture and Markov switching models. To the best of our knowledge, our study is the first one that employs multivariate count data regression models based on the negative binomial distribution in connection with the mixture and Markov switching approach. Although the use of these econometric models in our application was motivated by market microstructure theory and applied to time series of the number of buy and sell transactions of a sample of stocks traded on the New York Stock exchange, they can in principle also be used in applications from other fields that study time series of count data.

In our empirical application we showed, how these methods may be used to analyze the trading process in a generalized version of the basic sequential trade model of Easley, Kiefer, O'Hara, and Paperman (1996). We found, consistent with the theory, that the bivariate trading process appears to be governed by a three component mixture density. Other aspects of the basic model were rejected by our findings. In particular, we found evidence of unobserved heterogeneity among the population of active traders, which may be caused by differences in individual specific characteristics. Also, we found our results to be consistent with an asymmetric version of the sequential trade model, in which buyers and sellers have different trade arrival rates. The basic model assumes instead, that arrival rates of buyers and sellers are the same.

Furthermore, our results indicate that informational epochs may have variable length, while the basic model assumes that all information epochs have the same length, namely one trading day.

Our evidence thus describes an intradaily trading process, that operates in absence of information events for long periods of time, driven mainly by uninformed transaction demands. These periods of 'normal' trading are frequently disrupted. Such interruptions are typically short lived and appear to be caused by the occurrence of information events, during which a higher level of trading activity on one or even both sides of the market is observed. We also observe overshooting market reactions during such informational epochs, i.e. the trading process is characterized by frequent switches between the good and bad news regime rather than by a persistent stay in either one of the two information regimes. In addition, informational epochs exhibit a higher variability of the trading activity.

There are a number of possible extensions of our approach, which may be the subject of further research. We want to point out a few directions, in which subsequent research projects might head. In Chapter 3 we reviewed a number of alternative sequential trade models. From a purely technical point of view, most of these models differed from the basic model only with respect to the number of trading events considered, and our econometric framework can be used without modifications to estimate reduced form versions of these models as well. For example, it would be possible to employ our methods in order to include information on the volume of the trades, to discriminate between different order types, or to analyze parallel trading on different markets. Also, versions of the noise trader model introduced by Weston (2001) with more than three information regimes can be estimated employing our framework, although our empirical evidence suggests that for our sample of stocks mixture models based on three regimes yield the best fit.

In this study we used count data regression models based on the Poisson and the negative binomial density. The results of our specification tests suggest that the fit of our models to the data might be improved with respect to the distributional assumptions that we specified for the trade events. Therefore, a second line of departure for future studies might be to combine the mixture and Markov switching approach with other marginal densities for the trade events. In addition, the information content of 'no trade' intervals might be analyzed by employing count data models that allow one to model zero counts differently from other values that the dependent variables can assume by means of a hurdle regression model as in Mullahy (1986) or a zero-inflated count data regression model as in Lambert (1992).[1]

Another issue for future contributions is the estimation of structural models based on the EM-algorithm. As noted in Chapter 4, structural models have the disadvantage, that there is no obvious manner, in which the number of mixture components can be assessed consistently. Therefore, it might also

---

[1] See also Greene (1994).

be worthwhile to develop improved statistical tests for the number of regime components that could be used for that purpose.

An alternative approach for the analysis of sequential trade models based on intradaily sampling schemes has recently been suggested by Hujer, Vuletić, and Kokot (2002). They model the time series of durations between successive trades by means of a three regime Markov switching ACD approach, but their empirical model lacks a close connection to the economic theory. Their specification of the relationship between the regimes that are apparent for the time series of trade durations and the observable trade events is rather loose and can be motivated economically only on an ad hoc basis. Recently however, the ACD-framework has been extended to the multivariate case as well, see Russell and Engle (1999) and Russell (1999). A promising strategy for future research in this vein would be to combine their approach with a multivariate extension of the ACD model. This would allow one to develop more natural tests of the implications of sequential trade models on a trade by trade basis.

A last conclusion to be drawn from this study, is that financial transaction data sets should include the true trade direction if possible. It should not be too difficult or expensive to record this important piece of information, given the recent evolution of electronic trading systems. Such information should be made available to researchers. One could assess the magnitude of the misclassification bias. Alternatively, one could think of applying an 'errors in variables' approach to estimate sequential trade models based on possibly misclassified data sets. Such models have been proposed in the context of univariate count data models by Whittemore and Gong (1991). However, the implementation of such models in a mixture framework may turn out to be very complicated and is clearly beyond the scope of the present study. All in all we hope, that this contribution will encourage other researchers to improve and extend our approach even further.

# Appendix

## A.1 The Poisson Process

Consider analyzing a process of *events* that occur randomly in time.[1] An example is the arrival of customers at a service counter. A statistical model that may be used to describe this stochastic process is the Poisson process, which is based on the following four postulates:

1. The number of events in non-overlapping intervals of time are independent random variables (*independent increments*).
2. The probability distribution of the number of events is independent of the location in time (*stationarity*).
3. The probability of exactly one event in a small interval of time is approximately proportional to the length of the interval.
4. The probability of *more* than one event in a small interval of time is negligible.

The last two postulates of the Poisson process may be defined more precisely as follows. The counting process $N(\tau), \tau \geq 0$ is equal to the total number of events observed by time $\tau$. The random variable $N(\tau)$ is nonnegative and integer valued, and satisfies the property that $N(\tau_i) \leq N(\tau_j)$ if $\tau_i < \tau_j$. Events occur in time according to the following probability law:

$$p[N(\tau + \Delta\tau) - N(\tau) = 0] = 1 - \lambda \cdot \Delta\tau + o(\Delta\tau),$$
$$p[N(\tau + \Delta\tau) - N(\tau) = 1] = \lambda \cdot \Delta\tau + o(\Delta\tau),$$

which implies

$$p[N(\tau + \Delta\tau) - N(\tau) > 1] = o(\Delta\tau),$$

where the expression $o(x)$ denotes any function of $x$ such that $\lim_{x \to 0} \frac{o(x)}{x} = 0$.

---

[1] For a comprehensive review on Poisson and related stochastic processes consider e.g. Kingman (1993), or Ross (1996).

Thus, the probability of observing an event, such as a customer arrival in the interval $(\tau, \tau + \Delta\tau)$ depends on the length of the interval $\Delta\tau$ and on the *intensity* $\lambda$, which is the instantaneous rate of events that occur per unit of time. As long as postulates (1) - (4) hold, the discrete valued random variable

$$y(\Delta\tau) \equiv N(\tau + \Delta\tau) - N(\tau)$$
$$= \text{number of events in an interval of fixed length } \Delta\tau$$

is distributed according to the Poisson distribution[2], which has probability density function

$$f_P(y) = \frac{(\lambda \cdot \Delta\tau)^y}{y!} \cdot \exp(-\lambda \cdot \Delta\tau),$$

and cumulative density

$$F_P(y) = \sum_{i=1}^{y} \frac{(\lambda \cdot \Delta\tau)^i}{i!} \cdot \exp(-\lambda \cdot \Delta\tau).$$

Another random variable associated with the Poisson process is the waiting time $x(\tau)$ from one event to the next, which is a positive valued, continuous random variable. Let $\tau_0$ denote an arbitrary point in time at which a stopwatch is started. Let $x(\tau)$ denote the time span from $\tau_0$ to the first event thereafter (occurring at time $\tau > \tau_0$), so in general $x(\tau) \equiv \Delta\tau_0 = \tau - \tau_0$. The duration $x(\tau)$ will exceed any given $x > 0$ if and only if there are no events in the intervening interval $[\tau_0, \tau]$, so

$$p[x(\tau) > x] = p[N(\tau) - N(\tau_0) = 0]$$
$$= p[y(\tau - \tau_0) = 0]$$
$$= \frac{(\lambda \cdot x)^0}{0!} \cdot \exp(-\lambda \cdot x)$$
$$= \exp(-\lambda \cdot x),$$

for $x > 0$. The cumulative density function of $x$ is obtained by complementation

$$F_E(x) = p[x(\tau) < x]$$
$$= 1 - \exp(-\lambda \cdot x),$$

which is an exponential distribution with mean $E(x) = \frac{1}{\lambda}$ and variance $Var(x) = \frac{1}{\lambda^2}$. The corresponding density function of $x$ is equal to the derivative of the c.d.f. with respect to $x$ and is thus given by

$$f_E(x) = \frac{dF_E(x)}{dx}$$
$$= \lambda \cdot \exp(-\lambda \cdot x).$$

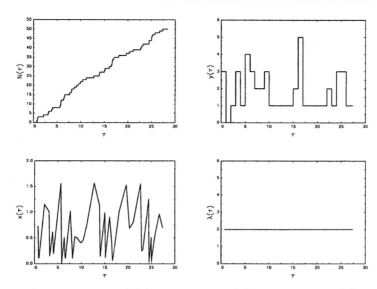

**Fig. A.1.** Counting process $N(\tau)$, event counts $y(\tau)$, waiting times $x(\tau)$ and intensity function $\lambda(\tau)$ for a simulated Poisson process with $\lambda = 2$.

Figure A.1 shows plots of the counting process $N(\tau)$, the number of events in a one unit time interval $y(\tau)$, the waiting times between events $x(\tau)$ and the associated intensity function $\lambda(\tau)$ from a simulated Poisson process, for which 50 independent random variables were drawn from an exponential distribution with $\lambda$ set equal to 2.

## A.2 Maximum Likelihood Estimation of a Multivariate Poisson Mixture Model

Maximum likelihood estimates for the parameters of the ignorant agnostic model may be found by forming the Lagrangean

$$J\left(\theta\right) = \ln \mathcal{L}_I\left(\theta\right) + \gamma\left(1 - \sum_{j=1}^{N} \pi_j\right),$$

and setting it's derivative with respect to $\theta$ equal to zero. Note that

$$\frac{\partial \ln \mathcal{L}_I\left(\theta\right)}{\partial \theta} = \sum_{t=1}^{T} \frac{1}{f_t\left(y_t; \theta\right)} \cdot \frac{\partial f_t\left(y_t; \theta\right)}{\partial \theta}.$$

---

[2] See Lindgren, McElbrath, and Berry (1978), pp. 105-106 for a simple proof.

From (4.4) we obtain

$$\frac{\partial f_t\,(y_t;\theta)}{\partial \lambda_{jk}} = \pi_j \cdot \frac{\partial f_j\,(y_t\mid r_t = j;\theta)}{\partial \lambda_{jk}}$$

$$= \pi_j \cdot \frac{\partial f_{y_{tk}}\,(y_{tk}\mid r_t = j;\theta)}{\partial \lambda_{jk}} \cdot \prod_{\substack{m=1\\m\neq k}}^{K} f_{y_{tk}}\,(y_{tm}\mid r_t = j;\theta),$$

and from (4.2)

$$\frac{\partial f_{y_{tk}}\,(y_{tk}\mid r_t = j;\theta)}{\partial \lambda_{jk}} = \Delta t \cdot f_{y_{tk}}\,(y_{tk}\mid r_t = j;\theta) \cdot \Big[y_{tk}\cdot(\lambda_{jk}\cdot\Delta t)^{-1} - 1\Big].$$

Thus, using the definition of the density of the regime variable $p_t\,(r_t = j\mid y_t;\theta)$ conditional on the observed data in (4.6), we get

$$\frac{\partial \ln \mathcal{L}_I\,(\theta)}{\partial \lambda_{jk}} = \Delta t \cdot \sum_{t=1}^{T} \frac{\pi_j \cdot f_j\,(y_t\mid r_t = j;\theta)}{f_t\,(y_t;\theta)} \cdot \Big[y_{tk}\cdot(\lambda_{jk}\cdot\Delta t)^{-1} - 1\Big] \quad \text{(A.1)}$$

$$= \Delta t \cdot \sum_{t=1}^{T} p_t\,(r_t = j\mid y_t;\theta) \cdot \Big[y_{tk}\cdot(\lambda_{jk}\cdot\Delta t)^{-1} - 1\Big]$$

$$= \Delta t \cdot \Bigg[(\lambda_{jk}\cdot\Delta t)^{-1} \cdot \sum_{t=1}^{T} y_{tk}\cdot p_t\,(r_t = j\mid y_t;\theta)$$

$$- \sum_{t=1}^{T} p_t\,(r_t = j\mid y_t;\theta)\Bigg].$$

The derivative of the log likelihood with respect to $\pi_j$ can be derived in an analogous manner. It is given by

$$\frac{\partial \ln \mathcal{L}_I\,(\theta)}{\partial \pi_j} = \pi_j^{-1} \cdot \sum_{t=1}^{T} p_t\,(r_t = j\mid y_t;\theta).$$

Setting the derivatives of the Lagrangean $J\,(\theta)$ with respect to $\lambda_{jk}$ equal to zero is equivalent to setting (A.1) equal to zero. The corresponding lines of (4.7) are obtained by rearranging these equations. In order to derive the corresponding equation for $\pi_j$, we start with

$$\frac{\partial J\,(\theta)}{\partial \pi_j} = \pi_j^{-1} \cdot \sum_{t=1}^{T} p_t\,(r_t = j\mid y_t;\theta) - \gamma \overset{!}{=} 0. \quad \text{(A.2)}$$

Rearranging and summing the expressions in (A.2) over all possible states $j = 1, 2, \ldots, N$ yields

$$\sum_{t=1}^{T} \sum_{j=1}^{N} p_t\,(r_t = j\mid y_t;\theta) = \gamma \cdot \sum_{j=1}^{N} \pi_j.$$

But

$$\sum_{j=1}^{N} p_t\left(r_t = j \mid y_t; \theta\right) = \sum_{j=1}^{N} \pi_j = 1$$

so that $\gamma = T$. Substituting $\gamma$ in (A.2) and solving for $\pi_j$ yields (4.8).

## A.3 The EM-Algorithm

In this Section we will provide an exposition of the basic principle that makes the EM-algorithm applicable in many situations. The general setting, in which the EM-algorithm may be applied to estimate parameters of interest, is when the data at hand is incomplete, i.e. when some variables or observations are not available in the data set used for estimation. The following exposition draws heavily on the work of Dempster, Laird, and Rubin (1977), and Ruud (1991).

We assume the existence of two sample spaces $\mathcal{Y}$, the sample space of the *observed* (*incomplete*) data $y$, and $\mathcal{W}$, the sample space of the *unobserved* (*complete*) data $w$, and a 'many to one' mapping from $\mathcal{W}$ to $\mathcal{Y}$. The observed data $y$ are obtained by a random draw from the sample space $\mathcal{Y}$, and the corresponding $w$ may only be observed indirectly through $y$, or to be more precisely, through the mapping $w \rightarrow y\left(w\right)$. Thus, $w$ is known only to lie in $\mathcal{W}\left(y\right)$, i.e. the subset of $\mathcal{W}$ determined by the equation $y = y\left(w\right)$. The mapping $w \rightarrow y\left(w\right)$ is often called the 'observation rule', and it should be understood, that it may not depend on parameters to be estimated in such a way that $\mathcal{W}\left(y\right)$ depends on these parameters.[3]

We also assume the existence of a family of sampling densities $f\left(w|\theta\right)$, for the complete data, and $g\left(y|\theta\right)$ for the incomplete data. Both depend on the same, $k$ dimensional set of parameters $\theta \in \Theta$, where $\Theta$ is a compact subset of $\mathbb{R}^k$. The relation between those two densities is given by

$$g\left(y|\theta\right) = \int_{\mathcal{W}(y)} f\left(w|\theta\right) dx, \tag{A.3}$$

i.e. the incomplete density $g(\cdot)$ is obtained from the complete density $f(\cdot)$ by 'integrating out' the unobserved data $w$ over its sample space $\mathcal{W}\left(y\right)$. An alternative point of view that is useful in many circumstances is to interpret $f\left(\cdot|\theta\right)$ as the *joint density* of the observable and the unobservable data and $g\left(\cdot|\theta\right)$ as the corresponding marginal density of the observable data. This implies that the observable data $y$ are a subset of the unobservable $w$ and the corresponding sample spaces are related by $\mathcal{Y} \subseteq \mathcal{W}$. In general, for a given incomplete data specification $g\left(\cdot|\theta\right)$, many possible complete data specifications $f\left(\cdot|\theta\right)$ may be defined, but often one of them appears to be a 'natural choice'.

---

[3] Otherwise the unimodality of the function $H(\cdot)$, defined below in (A.9), may not be established, see Ruud (1991), p. 310.

Now, let us denote the *complete likelihood function* of the parameter vector $\theta$ as

$$\mathcal{L}_C\left(\theta|W_T\right) = \prod_{t=1}^{T} f\left(w_t|\theta\right), \tag{A.4}$$

the associated *incomplete likelihood function* as

$$\mathcal{L}_I\left(\theta|Y_T\right) = \prod_{t=1}^{T} g\left(y_t|\theta\right), \tag{A.5}$$

where $T$ is the size of the sample at hand, $W_T$ is a $(T \times u)$ matrix of unobservable data, $Y_T$ is a $(T \times v)$ matrix of observable data, and $w_t$ respectively $y_t$ are the corresponding column vectors of observations at time $t$. When the observable data is a subset of the unobservable, we have $u > v$.[4] Associated with this statistical model is the conditional density of $w$, given $y$ and $\theta$, which is given by

$$k\left(w_t|\theta, y_t\right) = \frac{f\left(w_t|\theta\right)}{g\left(y_t|\theta\right)}. \tag{A.6}$$

and the *conditional complete likelihood function*

$$\mathcal{L}_K\left(\theta|Y_T\right) = \prod_{t=1}^{T} k\left(w_t|\theta, y_t\right). \tag{A.7}$$

In order to define the EM-algorithm we also need to introduce the *expected complete log likelihood function* as

$$Q\left(\theta|\theta^{(p)}, Y_T\right) = E\left[\ln \mathcal{L}_C\left(\theta|W_T\right)|\theta^{(p)}, Y_T\right]. \tag{A.8}$$

In the context of the EM-algorithm, $\theta^{(p)}$ is a provisional guess for the parameter vector $\theta$, which is used to evaluate the expectation of the complete log likelihood. A typical iteration of the EM-algorithm is then defined by two steps, the *expectation step* (E-step), in which the conditional expectation of the complete log likelihood given the observations $Y_T$ and the current guess of the parameter vector $\theta^{(p)}$ is formed and the *maximization step* (M-step), in which this conditional expectation is being maximized with respect to $\theta$ in order to obtain the next provisional guess $\theta^{(p+1)}$:

E-step:  Compute $Q\left(\theta|\theta^{(p)}, Y_T\right)$.

M-step:  Choose $\theta^{(p+1)}$ as the value of $\theta \in \Theta$ which maximizes $Q\left(\theta|\theta^{(p)}, Y_T\right)$, i.e.
$$\theta^{(p+1)} \equiv \max_{\theta \in \Theta} Q\left(\theta|\theta^{(p)}, Y_T\right).$$

---

[4] Actually, the application of the EM-algorithm does not necessarily require that the sample sizes of the observable and unobservable data are equal. The EM-algorithm may also be used in situations when all variables of the statistical model are observed, i.e. $u = v$, but some observations are missing, so that $W_T$ is of dimension $(T_1 \times u)$ and $Y_T$ has dimension $(T_2 \times u)$ with $T_1 > T_2$.

The intuition behind the EM-algorithm may be described as follows: We would like to choose an estimator $\hat{\theta}$ which maximizes the complete likelihood $\mathcal{L}_C(\theta|W_T)$. Since this estimator is infeasible given only observations on $y$, we instead maximize its expectation given the sample $Y_T$ and the current guess $\theta^{(p)}$. Convergence of the EM-algorithm is achieved when $\max | \theta^{(p+1)} - \theta^{(p)} | \leq \epsilon$, where the convergence criterion $\epsilon$ is set to a very small number in practice, say $10^{-5}$.

Under conditions described in Boyles (1983), Wu (1983) and Redner and Walker (1985) the iterative application of this algorithm yields a sequence of values for $\theta$ which converges to a local maximum of $\mathcal{L}_I(\theta|Y_T)$. The difference between the observed, incomplete log likelihood $\ln \mathcal{L}_I(\theta|Y_T)$ and $Q(\theta|\theta^{(p)}, Y_T)$, which is actually maximized, is equal to

$$
\begin{aligned}
H\left(\theta|\theta^{(p)}, Y_T\right) &= Q\left(\theta|\theta^{(p)}, Y_T\right) - \ln \mathcal{L}_I(\theta|Y_T) \\
&\equiv E\left[\ln \mathcal{L}_K(\theta|Y_T)|\theta^{(p)}, Y_T\right].
\end{aligned}
\tag{A.9}
$$

It can be shown that for any pair $\left(\theta^{(p+1)}, \theta^{(p)}\right)$ in $\Theta \times \Theta$

$$
H\left(\theta|\theta^{(p+1)}, Y_T\right) \leq H\left(\theta|\theta^{(p)}, Y_T\right),
\tag{A.10}
$$

with equality if and only if

$$
E\left[\ln \mathcal{L}_K(\theta|Y_T)|\theta^{(p+1)}, Y_T\right] = E\left[\ln \mathcal{L}_K(\theta|Y_T)|\theta^{(p)}, Y_T\right],
$$

and $\mathcal{L}_K(\theta|Y_T) > 0$ for all $w \in \mathcal{W}(y)$. A simple proof of this assertion is given in Hamilton (1990), pp. 48-49.

Generalized versions of the EM-algorithm may be defined by replacing the M-step as defined above by any iterative solution with associated mapping $\theta^{(p+1)} = M\left(\theta^{(p)}|Y_T\right)$, which yields a new guess $\theta^{(p+1)}$ that increases the value of $Q(\theta|\theta^{(p)}, Y_T)$. Thus, any iterative algorithm is a generalized EM-algorithm (GEM-algorithm) if it satisfies

$$
\begin{aligned}
Q\left(M\left(\theta^{(p+1)}|Y_T\right)|\theta^{(p+1)}, Y_T\right) &\geq Q\left(M\left(\theta^{(p)}|Y_T\right)|\theta^{(p)}, Y_T\right) \\
\Rightarrow \mathcal{L}_I\left(M\left(\theta^{(p+1)}|Y_T\right)|Y_T\right) &\geq \mathcal{L}_I\left(M\left(\theta^{(p)}|Y_T\right)|Y_T\right),
\end{aligned}
\tag{A.11}
$$

for every $\theta \in \Theta$. Note in particular, that it is not necessary to maximize $Q(\theta|\theta^{(p)}, Y_T)$ in every iteration in order to qualify as a GEM-algorithm according to (A.11). The notion of a GEM-algorithm may be used to define alternative algorithms that improve on the speed of convergence. All the properties stated in this Section for the EM-algorithm apply equally to any other algorithm that satisfies (A.11).

An estimator of the information matrix $\mathcal{I}(\theta)$ may be obtained in the context of the EM-algorithm by differentiating (A.9) with respect to $\theta$ and noting that, after convergence has been achieved, the derivative of $H\left(\theta|\theta^{(p)}, Y_T\right)$ is equal to zero. This implies

$$\left. \frac{\partial \ln \mathcal{L}_I(\theta|Y_T)}{\partial \theta} \right|_{\theta=\hat{\theta}} = \left. \frac{\partial Q(\theta|\theta^{(p)}, Y_T)}{\partial \theta} \right|_{\theta=\hat{\theta}}, \tag{A.12}$$

if $g(y|\theta)$ is differentiable with respect to $\theta$ and the mapping $y = y(w)$ does not depend on $\theta$. The proposed estimator of $\mathcal{I}(\theta)$ is

$$\hat{\mathcal{I}}_{OPG}(\theta) = T^{-1} \cdot \sum_{t=1}^{T} \left( \left. \frac{\partial Q(\theta|\theta^{(p)}, y_t)}{\partial \theta} \right|_{\theta=\hat{\theta}} \right) \cdot \left( \left. \frac{\partial Q(\theta|\theta^{(p)}, y_t)}{\partial \theta} \right|_{\theta=\hat{\theta}} \right)', \tag{A.13}$$

which is the EM-analog of the well known *outer product of gradients* (OPG) estimator of $\mathcal{I}(\theta)$ in maximum likelihood applications. An alternative estimator may be obtained as an analog of the information matrix estimator based on the Hessian matrix,

$$\hat{\mathcal{I}}_H(\theta) = -T^{-1} \cdot \left. \frac{\partial^2 Q(\theta|\theta^{(p)}, Y_T)}{\partial \theta \partial \theta'} \right|_{\theta=\hat{\theta}}, \tag{A.14}$$

which may be more easily computed by numerical differentiation of $Q(\theta|\theta^{(p)}, Y_T)$ with respect to $\theta$ in practice. A third estimator that would be the EM-analog of the *quasi maximum likelihood* (QML) estimator of the information matrix may be obtained by combining $\mathcal{I}_{OPG}(\theta)$ and $\mathcal{I}_H(\theta)$ as follows

$$\hat{\mathcal{I}}_{QML}(\theta) = \mathcal{I}_H(\theta) \cdot \mathcal{I}_{OPG}(\theta)^{-1} \cdot \mathcal{I}_H(\theta). \tag{A.15}$$

This estimator has been proposed by White (1982) in the context of ML-estimation when the density function of the observations is misspecified. He showed that in many situations estimates of the parameter vector $\theta$ may still be consistent despite of the misspecified density, but estimates of the information matrix computed either from (A.13) or (A.14) are biased and he supposed to use the consistent estimator $\hat{\mathcal{I}}_{QML}(\theta)$ instead. It should be noted however, that all three estimators of $\mathcal{I}(\theta)$ are only valid, if evaluated at the value of $\theta$ that maximizes the observed log likelihood, i.e. $\theta = \hat{\theta}$. Given either estimate of $\mathcal{I}(\theta)$, the computation of the covariance matrix of $\theta$ may precede as usual, i.e. $Cov(\hat{\theta}) = T^{-1} \cdot \hat{\mathcal{I}}(\hat{\theta})^{-1}$.

## A.4 The Poisson Regression Model

The basic regression model for count data is obtained by assuming, that the conditional distribution of the dependent variable $y_t$, given the regressor variables $x_t = (x_{t1}, \ldots, x_{tk})'$ is equal to the density of the Poisson distribution

$$f_P(y_t|x_t) = \frac{\exp(-\lambda_t \cdot \Delta t) \cdot (\lambda_t \cdot \Delta t)^{y_t}}{y_t!},$$

where $\Delta t$ is the length of the time interval during which the event counts $y_t$ are being observed, and $\lambda_t$ is the corresponding intensity i.e. it gives the expected number of events per unit of time. The conditional mean function of the count variable is specified as follows

$$E\left(y_t | x_t\right) = \lambda_t \cdot \Delta\, t = \exp\left(x_t' \beta\right) \cdot \Delta\, t,$$

and the conditional variance is equal to the conditional mean. Given a sample of $T$ independent observations on $y_t$ and $x_t$, the likelihood is given by

$$\mathcal{L}\left(\beta | y, X\right) = \prod_{t=1}^{T} f_P\left(y_t | x_t\right),$$

with $y = \left(y_1, \ldots, y_T\right)'$, the vector of observations on the dependent variable and $X = \left(x_1', \ldots, x_T'\right)$ the matrix of observations on the regressor variables. The corresponding log-likelihood is equal to

$$\ln \mathcal{L}\left(\beta | y, X\right) = \sum_{t=1}^{T} y_t \cdot x_t' \beta + y_t \cdot \ln \Delta\, t - \lambda_t \cdot \Delta\, t - \ln y_t!.$$

The gradient vector is given by

$$\frac{\partial \ln \mathcal{L}\left(\beta | y, X\right)}{\partial \beta} = \sum_{t=1}^{T} \left(y_t - \lambda_t \cdot \Delta\, t\right) \cdot x_t,$$

and the corresponding Hessian matrix is

$$\frac{\partial \ln \mathcal{L}\left(\beta | y, X\right)}{\partial \beta\, \partial \beta'} = -\sum_{t=1}^{T} \lambda_t \cdot \Delta\, t \cdot x_t x_t'.$$

The coefficients $\beta_i$ are to be interpreted as *semielasticities*, i.e. they give the proportionate change of the conditional mean, if the $i$-th regressor variable is changed by one unit

$$\beta_i = \frac{\partial\, \lambda_t \cdot \Delta\, t}{\partial x_{ti}} \cdot \frac{1}{\lambda_t \cdot \Delta\, t}.$$

If the $i$-th regressor enters logarithmically in the conditional mean specification, then $\beta_i$ is an *elasticity*, i.e. it gives the percentage change of the conditional mean, when the $i$-th regressor variable changes by 1%, see Cameron and Trivedi (1998), pp. 80-81.

## A.5 The Negative Binomial Regression Model

As in Sect. A.4, we assume that the mean parameter of the regression model $\lambda_t$ is determined by the log-linear specification

$$E\left(y_t | x_t\right) = \lambda_t \cdot \Delta\, t = \exp\left(x_t' \beta\right) \cdot \Delta\, t,$$

while the conditional density of the dependent variable $y_t$, given the regressor variables $x_t = (x_{t1}, \ldots, x_{tk})'$ is given by the negative binomial regression model known as Negbin 2 [see Cameron and Trivedi (1986) and Lawless (1987)] with density given by (4.20)

$$f_{NB2}(y_t \mid x_t; \beta, \alpha) = \frac{\Gamma\left(y_t + \alpha^{-1}\right)}{\Gamma\left(y_t + 1\right) \cdot \Gamma\left(\alpha^{-1}\right)}$$

$$\cdot \left(\frac{\alpha^{-1}}{\alpha^{-1} + \Delta t \cdot \lambda_t}\right)^{\alpha^{-1}} \cdot \left(\frac{\Delta t \cdot \lambda_t}{\alpha^{-1} + \Delta t \cdot \lambda_t}\right)^{y_t},$$

where $\Gamma(s) = \int_0^\infty \exp(-u) \cdot u^{s-1} du$ denotes the Gamma function.

The negative binomial density may be given a simpler form avoiding the evaluation of $\Gamma(\cdot)$ by exploiting the following property of the Gamma function

$$\Gamma(s) = \begin{cases} (s-1)\,\Gamma(s-1) & \text{if} \quad s > 0 \\ (s-1)! & \text{if} \quad s \in \mathbb{N}_0 \end{cases},$$

see Cameron and Trivedi (1998), p. 374. Thus, we may replace $\Gamma(y_t + 1)$ by $y_t!$, and

$$\frac{\Gamma\left(y_t + \alpha^{-1}\right)}{\Gamma\left(\alpha^{-1}\right)} = \begin{cases} \prod_{j=0}^{y_t - 1} \left(j + \alpha^{-1}\right), & \text{if} \quad y_t > 0 \\ 1, & \text{if} \quad y_t = 0 \end{cases}.$$

The negative binomial density can therefore be written as

$$f_{NB2}(y_t \mid x_t; \beta, \alpha) = \frac{1}{y_t!} \cdot \left(\frac{1}{1 + \alpha \cdot \lambda_t \cdot \Delta t}\right)^{\frac{1}{\alpha}} \cdot \left(\frac{\alpha \cdot \lambda_t \cdot \Delta t}{1 + \alpha \cdot \lambda_t \cdot \Delta t}\right)^{y_t}$$

$$\cdot \prod_{j=0}^{y_t - 1} \left(j + \frac{1}{\alpha}\right).$$

The Negbin 2 specification is appropriate for modelling *overdispersed* counts, which means that the conditional variance is greater than the conditional mean. The conditional variance function of the Negbin 2 density is given by $\lambda_t \cdot \Delta t + \alpha \cdot (\lambda_t \cdot \Delta t)^2$ instead of $\lambda_t \cdot \Delta t$ in the equidispersed Poisson model. Given a sample of $T$ independent observations on $y_t$ and $x_t$, the likelihood function of the parameter vector $\theta = (\beta_1, \ldots, \beta_k, \alpha)'$ is given by

$$\mathcal{L}(\theta \mid y, X) = \prod_{t=1}^{T} f_{NB2}(y_t \mid x_t),$$

with associated log likelihood

$$\ln \mathcal{L}(\theta \mid y, X) = \sum_{t=1}^{T} \left[ -\ln y_t! + \frac{1}{\alpha} \cdot \ln \left(\frac{1}{1 + \alpha \cdot \lambda_t \cdot \Delta t}\right) \right.$$

$$+ y_t \cdot \ln \left( \frac{1}{1 + \alpha \cdot \lambda_t \cdot \Delta t} \right) + y_t \cdot \ln \left( \alpha \cdot \lambda_t \cdot \Delta t \right)$$

$$+ \sum_{j=0}^{y_t - 1} \ln \left( j + \frac{1}{\alpha} \right) \Bigg]$$

$$= \sum_{t=1}^{T} \left[ - \ln y_t! - \left( \frac{1}{\alpha} + y_t \right) \cdot \ln \left( 1 + \alpha \cdot \lambda_t \cdot \Delta t \right) \right.$$

$$\left. + y_t \cdot \ln \left( \lambda_t \cdot \Delta t \right) + y_t \cdot \ln \alpha + \sum_{j=0}^{y_t - 1} \ln \left( j + \frac{1}{\alpha} \right) \right].$$

Since

$$y_t \cdot \ln \alpha + \sum_{j=0}^{y_t - 1} \ln \left( j + \frac{1}{\alpha} \right) = y_t \cdot \ln \alpha - \sum_{j=0}^{y_t - 1} \ln \left( \alpha \right) + \sum_{j=0}^{y_t - 1} \ln \left( j \cdot \alpha + 1 \right)$$

$$= \sum_{j=0}^{y_t - 1} \ln \left( j \cdot \alpha + 1 \right),$$

the log likelihood can be rewritten as

$$\ln \mathcal{L} \left( \theta | y, X \right) = \sum_{t=1}^{T} \left[ - \ln y_t! - \left( \frac{1}{\alpha} + y_t \right) \cdot \ln \left( 1 + \alpha \cdot \lambda_t \cdot \Delta t \right) \right.$$

$$\left. + y_t \cdot x_t' \beta + y_t \cdot \ln \Delta t + \sum_{j=0}^{y_t - 1} \ln \left( j \cdot \alpha + 1 \right) \right].$$

The derivative of the log likelihood function with respect to the regression parameter vector $\beta$ is given by

$$\frac{\partial \ln \mathcal{L} \left( \theta | y, X \right)}{\partial \beta} = \sum_{t=1}^{T} \left( \frac{y_t - \lambda_t \cdot \Delta t}{1 + \alpha \cdot \lambda_t \cdot \Delta t} \right) \cdot x_t,$$

and the derivative with respect to $\alpha$ is

$$\frac{\partial \ln \mathcal{L} \left( \theta | y, X \right)}{\partial \alpha} = \sum_{t=1}^{T} \left[ \frac{1}{\alpha^2} \ln \left( 1 + \alpha \cdot \lambda_t \cdot \Delta t \right) + \sum_{j=0}^{y_t - 1} \frac{j}{j \cdot \alpha + 1} \right.$$

$$\left. - \frac{\lambda_t \cdot \Delta t \cdot \left( y_t + \frac{1}{\alpha} \right)}{1 + \alpha \cdot \lambda_t \cdot \Delta t} \right].$$

The second derivative of the log likelihood with respect to the regression parameters $\beta$ is given by

$$\frac{\partial^2 \ln \mathcal{L}(\theta|y,X)}{\partial \beta \, \partial \beta'} = -\sum_{t=1}^{T} \frac{(1+\alpha \cdot y_t) \cdot \lambda_t \cdot \Delta t}{(1+\alpha \cdot \lambda_t \cdot \Delta t)^2} \cdot x_t \cdot x_t',$$

while the second derivative with respect to $\alpha$ is

$$\frac{\partial^2 \ln \mathcal{L}(\theta|y,X)}{\partial \alpha^2} = -\sum_{t=1}^{T} \left[ \sum_{j=0}^{y_t-1} \left( \frac{j}{j \cdot \alpha + 1} \right)^2 + \frac{2}{\alpha^3} \cdot \ln(1 + \alpha \cdot \lambda_t \cdot \Delta t) \right.$$
$$\left. - \frac{2\lambda_t \cdot \Delta t}{\alpha^2 \cdot (1 + \alpha \cdot \lambda_t \cdot \Delta t)} - \frac{(y_t + \frac{1}{\alpha}) \cdot (\lambda_t \cdot \Delta t)^2}{(1 + \alpha \cdot \lambda_t \cdot \Delta t)^2} \right].$$

The cross derivative is given by

$$\frac{\partial^2 \ln \mathcal{L}(\theta|y,X)}{\partial \beta \, \partial \alpha} = -\sum_{t=1}^{T} \frac{(y_t - \lambda_t \cdot \Delta t) \cdot \lambda_t \cdot \Delta t}{(1 + \alpha \cdot \lambda_t \cdot \Delta t)^2} \cdot x_t.$$

Note that for $\alpha \to 0$ the Negbin 2 model collapses to the ordinary Poisson regression model, see Cameron and Trivedi (1998), p. 75.

## A.6 Moments of Mixture Distributions

For finite mixture distributions with density given by[5]

$$f(y) = \sum_{j=1}^{N} \pi_j \cdot f_j(y \mid r = j),\,^{6}$$

the raw (uncentered) moments $\mu_s'$ are given by[7]

$$\mu_s' = E(y^s) = \sum_{j=1}^{N} \pi_j \cdot E(y^s \mid r = j). \tag{A.16}$$

The conditional raw moments $E(y^s \mid r = j)$ are equal to the corresponding expressions of the conditional distribution of the observable, dependent variable, given the unobservable regime

$$E(y^s \mid r = j) = \sum_{y=0}^{\infty} y^s \cdot f_j(y \mid r = j).$$

---

[5] In order to lessen the notational burden, we will neglect the time index $t$ in the following derivations and assume that the length of the observation interval $\Delta t$ is equal to one.

[6] This corresponds to the model without covariates presented in Sect. 4.1.2.

[7] See Cameron and Trivedi (1998), p. 130.

Assuming that $y$ has a conditional Poisson distribution yields[8]

$$E\left(y \mid r = j\right) = \lambda_j,$$
$$E\left(y^2 \mid r = j\right) = \lambda_j + \lambda_j^2.$$

Thus the mean of a mixture of Poisson distributions is given by

$$E\left(y\right) = \sum_{j=1}^{N} \pi_j \cdot \lambda_j \qquad (A.17)$$

and the second raw moment is

$$E\left(y^2\right) = \sum_{j=1}^{N} \pi_j \cdot \left(\lambda_j + \lambda_j^2\right).$$

The variance of the mixture may be calculated from the raw moments using the standard formula

$$Var\left(y\right) = E\left(y^2\right) - \left[E\left(y\right)\right]^2.$$

This yields the following expression for the variance of the Poisson mixture

$$Var\left(y\right) = \sum_{j=1}^{N} \pi_j \cdot \lambda_j + \sum_{j=1}^{N} \pi_j \cdot \lambda_j^2 - \left[\sum_{j=1}^{N} \pi_j \cdot \lambda_j\right]^2. \qquad (A.18)$$

Note that this may be rewritten as

$$Var\left(y\right) = E\left[Var\left(y \mid r = j\right)\right] + Var\left[E\left(y \mid r = j\right)\right]. \qquad (A.19)$$

The first term follows from the equality of mean and variance, which is a property of the Poisson distribution, so $Var\left(y \mid r = j\right) = E\left(y \mid r = j\right)$ and

$$E\left[E\left(y \mid r = j\right)\right] = E\left(y\right) = \sum_{j=1}^{N} \pi_j \cdot \lambda_j,$$

by the law of iterated expectations. The second term may be derived by noting that

$$Var\left[E\left(y \mid r = j\right)\right] = E\left(\lambda_j^2\right) - \left[E\left(\lambda_j\right)\right]^2$$

---

[8] These expressions may be derived from the moment generating function of the Poisson distribution, which is given by $E\left[\exp\left(q \cdot y\right)\right] = \exp\left[\lambda \cdot \left(\exp\left(q\right) - 1\right)\right]$. Differentiating the moment generating function $s$ times with respect to $q$ and evaluating it at $q = 0$ yields the $s$-th raw moment $E\left(y^s\right)$. See Johnson and Kotz (1969), p. 90.

$$= \sum_{j=1}^{N} p\left(r = j\right) \cdot \lambda_j^2 - \left[\sum_{j=1}^{N} p\left(r = j\right) \cdot \lambda_j\right]^2$$

$$= \sum_{j=1}^{N} \pi_j \cdot \lambda_j^2 - \left[\sum_{j=1}^{N} \pi_j \cdot \lambda_j\right]^2 .$$

Therefore, the variance of $y$ will exceed its mean, whenever the second term in (A.19) is greater than zero, i.e. whenever there is some variation among the conditional means $\lambda_j$. Hence, a Poisson mixture is able to cope with overdispersion.[9] The overdispersion property of the mixture density is also visible in a plot of the density function, since it contains more probability mass in its tails, than the corresponding one component density with the same unconditional mean. A mixture of Poisson densities can even cope with multimodality, see Fig. A.2 for an example.

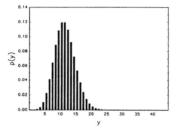 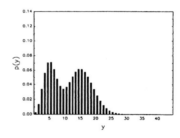

**Fig. A.2.** Density function of the Poisson distribution. Left panel: One component Poisson density with $\lambda = 11$. Right panel: Mixture of two Poisson densities with $\lambda_1 = 5$, $\lambda_2 = 15$ and $\pi_1 = 0.4$.

If the conditional distribution of $y$ is of the Negbin 2-type with parameters $\lambda$ and $\alpha$, the mean is again given by

$$E\left(y\right) = \sum_{j=1}^{N} \pi_j \cdot \lambda_j. \tag{A.20}$$

---

[9] For computational purposes it is preferable to rewrite the variance of the conditional means as a weighted sum of squared differences

$$Var\left[E\left(y \mid r = j\right)\right] = E\left[\lambda_j - E\left(\lambda_j\right)\right]^2$$
$$= \sum_{j=1}^{N} \pi_j \cdot \left(\lambda_j - \sum_{j=1}^{N} \pi_j \cdot \lambda_j\right)^2 .$$

From the moment generating function of the negative binomial distribution we get the following expression for the second raw moment[10]

$$E(y^2 \mid r = j) = \lambda_j + \alpha_j \lambda_j^2 + \lambda_j^2,$$

so the variance of a Negbin 2 mixture is given by

$$Var\,(y) = \sum_{j=1}^{N} \pi_j \cdot \left(\lambda_j + \alpha_j \lambda_j^2\right) + \sum_{j=1}^{N} \pi_j \cdot \lambda_j^2 - \left[\sum_{j=1}^{N} \pi_j \cdot \lambda_j\right]^2. \qquad (A.21)$$

Recalling that the conditional variance of the negative binomial distribution is given by

$$Var\,(y \mid r = j) = \lambda_j + \alpha_j \lambda_j^2,$$

we may immediately conclude, that the unconditional variance is again of the form (A.19). Thus, the overdispersion coefficient of the Negbin 2 mixture is greater than or equal to the corresponding coefficient of a one component Negbin model, with equality occurring when there is no variation in the conditional means $\lambda_j$. This is again visible in a comparison of the plots of the one component density function to the mixture density, see Fig. A.3. Higher moments of any desired order may be derived in the same manner, provided that the corresponding moments of the conditional distributions $f_j(y \mid r = j)$ exist.

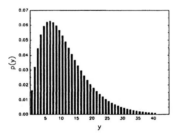 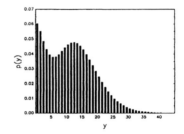

**Fig. A.3.** Density function of the negative binomial distribution. Left panel: One component negative binomial density with $\lambda = 11$ and $\alpha = 0.42$. Right panel: Mixture of two negative binomial densities with $\lambda_1 = 5$, $\alpha_1 = 0.9$, $\lambda_2 = 15$, $\alpha_2 = 0.1$, and $\pi_1 = 0.4$.

Next, let us consider the multivariate case. From the unconditional joint density of the vector $y = (y_1, \ldots, y_K)'$ which was given in (4.4), the unconditional marginal distribution of $y_s$ can be determined, by summing the

---

[10] The moment generating function for the Negbin 2 parameterization is given by $E\,[\exp\,(y \cdot q)] = [\lambda \alpha + 1 - \lambda \alpha \cdot \exp\,(q)]^{-\alpha^{-1}}$, see Johnson and Kotz (1969), p. 125. Note, that they give the expression of the moment generating function in terms of the parameters $N = \alpha^{-1}$ and $P = \lambda \cdot \alpha$.

remaining $k - 1$ variables 'out'

$$f_{y_s}(y_s) = \sum_{y_1=0}^{\infty} \cdots \sum_{y_{s-1}=0}^{\infty} \sum_{y_{s+1}=0}^{\infty} \cdots \sum_{y_K=0}^{\infty} \sum_{j=1}^{N} \pi_j \cdot \prod_{k=1}^{K} f_{y_k}(y_k \mid r = j)$$

$$= \sum_{y_1=0}^{\infty} \cdots \sum_{y_K=0}^{\infty} \left[ \pi_1 \cdot f_{y_1}(y_1 \mid r = 1) \cdot \ldots \cdot f_{y_K}(y_K \mid r = 1) \right.$$

$$\left. + \ldots + \pi_N \cdot f_{y_1}(y_1 \mid r = N) \cdot \ldots \cdot f_{y_K}(y_K \mid r = N) \right]$$

$$= \pi_1 \cdot \sum_{y_1=0}^{\infty} f_{y_1}(y_1 \mid r = 1) \cdot \ldots \cdot \sum_{y_{s-1}=0}^{\infty} f_{y_{s-1}}(y_{s-1} \mid r = 1)$$

$$\cdot f_{y_s}(y_s \mid r = 1) \cdot \sum_{y_{s+1}=0}^{\infty} f_{y_{s+1}}(y_{s+1} \mid r = 1)$$

$$\cdot \ldots \cdot \sum_{y_K=0}^{\infty} f_{y_K}(y_K \mid r = 1) + \ldots + \pi_N \cdot \sum_{y_1=0}^{\infty} f_{y_1}(y_1 \mid r = N)$$

$$\cdot \ldots \cdot \sum_{y_{s-1}=0}^{\infty} f_{y_{s-1}}(y_{s-1} \mid r = N) \cdot f_{y_s}(y_s \mid r = N)$$

$$\cdot \sum_{y_{s+1}=0}^{\infty} f_{y_{s+1}}(y_{s+1} \mid r = N) \cdot \ldots \cdot \sum_{y_K=0}^{\infty} f_{y_K}(y_K \mid r = N)$$

$$= \sum_{j=1}^{N} \pi_j \cdot f_{y_s}(y_s \mid r = j),$$

so each of the $K$ unconditional marginal densities is equal to its univariate counterpart, and thus means and variances can be computed from the formulas given earlier for this case. Although we assumed conditional independence between the trading events, the model is consistent with contemporaneous correlation between the elements of the vector $y$ since all of them depend on the same (unobservable) state of the regime variable $r$, and it is through this link that either positive or negative correlation may be modelled in this framework.

In order to compute unconditional covariances, we examine pairwise relations. Starting from the standard formula

$$Cov[y_s, y_u] = E[y_s \cdot y_u] - E[y_s] \cdot E[y_u]$$

the product moment $E[y_s \cdot y_u]$ is given by

$$E[y_s \cdot y_u] = \sum_{y_s=0}^{\infty} \sum_{y_u=0}^{\infty} y_s \cdot y_u \cdot f(y_s \cap y_u)$$

$$= \sum_{y_s=0}^{\infty} \sum_{y_u=0}^{\infty} y_s \cdot y_u \cdot \sum_{j=1}^{N} \pi_j \cdot f_{y_s}(y_s \mid r = j) \cdot f_{y_u}(y_u \mid r = j)$$

$$= \pi_1 \cdot \sum_{y_s=0}^{\infty} y_s \cdot f_{y_s}(y_s \mid r = 1) \cdot \sum_{y_u=0}^{\infty} y_u \cdot f_{y_u}(y_u \mid r = 1)$$

$$+ \ldots + \pi_N \cdot \sum_{y_s=0}^{\infty} y_s \cdot f_{y_s}(y_s \mid r = N) \cdot \sum_{y_u=0}^{\infty} y_u \cdot f_{y_u}(y_u \mid r = N)$$

$$= \sum_{j=1}^{N} \pi_j \cdot \lambda_{js} \cdot \lambda_{ju},$$

when $s \neq u$ and

$$E[y_s \cdot y_u] = \sum_{j=1}^{N} \pi_j \cdot E\left(y^2 \mid r = j\right)$$

when $s = u$. This yields the following expression for the covariance between $y_s$ and $y_u$

$$Cov[y_s, y_u] = \sum_{j=1}^{N} \pi_j \cdot \lambda_{js} \cdot \lambda_{ju} - \sum_{j=1}^{N} \pi_j \cdot \lambda_{js} \cdot \sum_{j=1}^{N} \pi_j \cdot \lambda_{ju} \qquad (A.22)$$

$$= \sum_{j=1}^{N} \pi_j \cdot \lambda_{js} \cdot \lambda_{ju} - \sum_{j=1}^{N} \sum_{i=1}^{N} \pi_j \cdot \pi_i \cdot \lambda_{js} \cdot \lambda_{iu}$$

$$= \sum_{j=1}^{N} \pi_j \cdot \lambda_{js} \cdot \lambda_{ju} - \sum_{j=1}^{N} \pi_j^2 \cdot \lambda_{js} \cdot \lambda_{ju} - \sum_{j=1}^{N} \sum_{\substack{i=1 \\ i \neq j}}^{N} \pi_j \cdot \pi_i \cdot \lambda_{js} \cdot \lambda_{iu}$$

$$= \sum_{j=1}^{N} \pi_j \cdot (1 - \pi_j) \cdot \lambda_{js} \cdot \lambda_{ju} - \sum_{j=1}^{N} \sum_{\substack{i=1 \\ i \neq j}}^{N} \pi_j \cdot \pi_i \cdot \lambda_{js} \cdot \lambda_{iu},$$

when $s \neq u$ and the well known expression (A.19) for the variance, when $s = u$. From the first line of (A.22) we see that $y_s$ and $y_u$ will be uncorrelated with each other, whenever

$$E(y_u) = \frac{\sum\limits_{j=1}^{N} \pi_j \cdot \lambda_{js} \cdot \lambda_{ju}}{\sum\limits_{j=1}^{N} \pi_j \cdot \lambda_{js}},$$

which will always be the case, if

$$\lambda_{ju} = E(y_u) \quad \forall j \in (1, \ldots, N),$$

i.e. if the conditional mean of at least one of the variables does not vary with the state of the regime variable $r$. Note, that in the Poisson model this is equivalent to $y_u$ having a one component distribution. In the Negbin 2 model uncorrelatedness may arise, even if both variables come from a non-degenerate joint mixture distribution, because if the conditional means $\lambda_j$ do not vary across regimes the conditional variance parameters $\alpha_j$ may still be regime dependent. This will generate dependence between higher moments of the variables. From the last line of (A.22) we see, that negative correlations will only be observed, if a probability weighted measure of the co-movement of the conditional means of the two variables across different regimes (the cross terms in the second sum) is bigger than the probability weighted measure of the co-movement in the same regime (the terms of the first sum), i.e. the variables $y_s$ and $y_u$ tend to move in different directions when they are in the same regime and in the same direction, when in different regimes.

Predictions of the dependent variables in a regression framework may be obtained in the same way in both models. Conditioning on both, the regime variable $r$ and the exogenous variables $x$ the expectation of the number of trading events $y_s$ is equal to

$$E\left(y_s \mid r = m, x; \theta\right) = \sum_{y_s=0}^{\infty} y_s \cdot \sum_{j=1}^{N} I\left(r = j \mid \theta\right) \cdot f_{y_s}\left(y_s \mid r = j, x; \theta\right) \quad (A.23)$$

$$= \sum_{y_s=0}^{\infty} y_s \cdot f_{y_s}\left(y_s \mid r = m, x; \theta\right)$$

$$= \lambda_{ms}.$$

Predictions, that are not conditioned on the unobservable regime variable are given by

$$E\left(y_s \mid x; \theta\right) = \sum_{y_s=0}^{\infty} y_s \cdot f_{y_s}\left(y_s \mid x; \theta\right) \quad (A.24)$$

$$= \sum_{y_s=0}^{\infty} y_s \cdot \sum_{j=1}^{N} p\left(r = j \mid \theta\right) \cdot f_{y_s}\left(y_s \mid r = j, x; \theta\right)$$

$$= \sum_{y_s=0}^{\infty} \left[ y_s \cdot p\left(r = 1 \mid \theta\right) \cdot f_{y_s}\left(y_s \mid r = 1, x; \theta\right) + \ldots \right.$$

$$\left. + y_s \cdot p\left(r = N \mid \theta\right) \cdot f_{y_s}\left(y_s \mid r = N, x; \theta\right) \right]$$

$$= p\left(r = 1 \mid \theta\right) \cdot \sum_{y_s=0}^{\infty} y_s \cdot f_{y_s}\left(y_s \mid r = 1, x; \theta\right) + \ldots$$

$$+ p\left(r = N \mid \theta\right) \cdot \sum_{y_s=0}^{\infty} y_s \cdot f_{y_s}\left(y_s \mid r = N, x; \theta\right)$$

$$= \sum_{j=1}^{N} p\,(r = j \mid \theta) \cdot \sum_{y_s=0}^{\infty} y_s \cdot f_{y_s}\,(y_s \mid r = j, x; \theta)$$

$$= \sum_{j=1}^{N} \pi_j \cdot \lambda_{jk},$$

since $r$ is assumed to be independent of $x$ and $p(r = j \mid \theta) = \pi_j$. Expressions for the variances and covariances in the regression framework as well as for higher moments can be derived in analogous manner.

## A.7 Unobserved Individual Variation of Trade Arrival Rates

We assume that the arrival rate of the $k$-th trading event under regime $j$ for the $s$-th trader is a random variable given by[11]

$$\lambda_{sjk} = \exp(x'\beta_{jk} + \nu_{sjk})$$
$$= \exp(x'\beta_{jk}) \cdot \exp(\nu_{sjk})$$
$$= \lambda_{jk} \cdot \psi_{sjk},$$

where $\lambda_{jk} \equiv \exp(x'\beta_{jk})$ is the mean arrival rate for the $k$-th trading event under regime $j$, which depends on observable quantities contained in the vector $x$ only, $\nu_{sjk}$ is an individual specific random effect that may depend on the unobservable characteristics and preferences of the $s$-th trader, and $\psi_{sjk} \equiv \exp(\nu_{sjk})$ represents the corresponding multiplicative shift of the arrival rate. $\psi_{sjk}$ is a continuous, non-negative random variable defined on $\mathbb{R}_+$. The density of the $k$-th trading event $y_k$ conditional on $\lambda_{sjk}$ is of the Poisson form

$$f(y_k|\lambda_{sjk}) = \frac{\exp(-\lambda_{sjk}) \cdot (\lambda_{sjk})^{y_k}}{y_k!},$$

while the unconditional density $f(y_k|\lambda_{jk})$ is obtained by integrating $\psi_{sjk}$ out

$$f(y_k|\lambda_{jk}) = \int_{0}^{\infty} f(y_k|\lambda_{sjk}) \cdot h(\psi_{sjk})\, d\psi_{sjk}. \tag{A.25}$$

If we assume that $\psi_{sjk}$ follows a Gamma distribution with mean equal to one and variance equal to $\alpha_{jk}$ and thus has density function $h(\psi_{sjk})$ given by

---

[11] For notational simplification we neglect the time index $t$ in the following derivation and assume that $\Delta t = 1$. A collection of alternative chance mechanisms that generate a negative binomial distribution may be found in Boswell and Patil (1970). See also Cameron and Trivedi (1998), pp. 100-102.

$$h(\psi_{sjk}) = \frac{(\alpha_{jk}^{-1})^{\alpha_{jk}^{-1}}}{\Gamma(\alpha_{jk}^{-1})} \cdot (\psi_{sjk})^{(\alpha_{jk}^{-1}-1)} \cdot \exp(-\psi_{sjk} \cdot \alpha_{jk}^{-1}), \qquad (A.26)$$

then the integral (A.25) may be expressed in a closed form. Replacing the random variable $\psi_{sjk}$ in (A.26) by $\frac{\lambda_{sjk}}{\lambda_{jk}}$, and noting that the Jacobian term of the transformation from $\psi_{sjk}$ to $\lambda_{sjk}$ is equal to $(\lambda_{jk})^{-1}$, the unconditional density may be written as

$$f(y_k|\lambda_{jk}) = \int_0^\infty \left[ \frac{\exp(-\lambda_{sjk}) \cdot (\lambda_{sjk})^{y_k}}{y_k!} \cdot \frac{(\frac{\alpha_{jk}^{-1}}{\lambda_{jk}})^{\alpha_{jk}^{-1}}}{\Gamma(\alpha_{jk}^{-1})} \right. \qquad (A.27)$$

$$\left. \cdot (\lambda_{sjk})^{(\alpha_{jk}^{-1}-1)} \cdot \exp\left( -\frac{\alpha_{jk}^{-1}}{\lambda_{jk}} \cdot \lambda_{sjk} \right) \right] d\lambda_{sjk}$$

$$= \frac{(\frac{\alpha_{jk}^{-1}}{\lambda_{jk}})^{\alpha_{jk}^{-1}}}{\Gamma(\alpha_{jk}^{-1}) \cdot \Gamma(y_k + 1)}$$

$$\cdot \int_0^\infty \left[ \exp\left( -\lambda_{sjk} \cdot \left( 1 + \frac{\lambda_{jk}}{\alpha_{jk}^{-1}} \right) \right) \cdot (\lambda_{sjk})^{(y_k + \alpha_{jk}^{-1} - 1)} \right] d\lambda_{sjk},$$

since $y_k! = \Gamma(y_k + 1)$. Noting, that the Gamma function satisfies the following property[12]

$$\Gamma(a) = b^a \cdot \int_0^\infty \exp(-u \cdot b) \cdot u^{a-1} \, du,$$

we can solve the integral appearing in the last line of (A.27) by setting $u \equiv \lambda_{sjk}$, $b \equiv 1 + \frac{\lambda_{jk}}{\alpha_{jk}^{-1}}$, and $a \equiv y_k + \alpha_{jk}^{-1}$, and obtain

$$f(y_k|\lambda_{jk}) = \frac{\Gamma(\alpha_{jk}^{-1} + y_k)}{\Gamma(\alpha_{jk}^{-1}) \cdot \Gamma(y_k + 1)} \cdot \left( \frac{\alpha_{jk}^{-1}}{\lambda_{jk}} \right)^{\alpha_{jk}^{-1}} \cdot \left( 1 + \frac{\alpha_{jk}^{-1}}{\lambda_{jk}} \right)^{-(\alpha_{jk}^{-1} + y_k)} \qquad (A.28)$$

$$= \frac{\Gamma(\alpha_{jk}^{-1} + y_k)}{\Gamma(\alpha_{jk}^{-1}) \cdot \Gamma(y_k + 1)} \cdot \left( \frac{\alpha_{jk}^{-1}}{\alpha_{jk}^{-1} + \lambda_{jk}} \right)^{\alpha_{jk}^{-1}} \cdot \left( \frac{\lambda_{jk}}{\alpha_{jk}^{-1} + \lambda_{jk}} \right)^{y_k},$$

which is a Negbin 2 density function with mean equal to $\lambda_{jk}$ and conditional variance equal to $\lambda_{jk} + \alpha_{jk} \cdot (\lambda_{jk})^2$.

---

[12] See Abramowitz and Stegun (1965), p. 255.

# A.8 Markov Chains

In this Section, we will summarize some of the basic properties of Markov chains that will be useful in this study.[13] Consider observing a time series of a random variable $r_t$ that can assume only integer values in the finite range $(1, 2, \ldots, N)$. If the conditional probability that $r_t$ assumes some particular value $j$ depends on the history of the process $r_{t-1}, r_{t-2}, \ldots$ only through its most recent realization $r_{t-1}$, i.e.

$$p(r_t = j \mid r_{t-1} = i, r_{t-2} = k, \ldots) = p(r_t = j \mid r_{t-1} = i) \equiv p_{ji},$$

this stochastic process is called a *homogenous N-state Markov chain*. Markov chains are the most fundamental time series models for qualitative and discrete-valued random variables.

The coefficients $p_{ji}$ are called *transition probabilities* since they represent the probability that state $i$ at time $t - 1$ will be followed by state $j$ at time $t$. A convenient representation of a Markov chain is obtained by collecting the probabilities $p_{ji}$ in a $(N \times N)$ *transition probability matrix* $P$ that relates the state of the process in the current period $t$ to the state of the process in past periods. $P$ can in general be be written as

$$P = \begin{bmatrix} p_{11} & p_{12} & \cdots & p_{1N} \\ \vdots & \vdots & \cdots & \vdots \\ p_{N1} & p_{N2} & \cdots & p_{NN} \end{bmatrix}.$$

The elements of $P$ are all non-negative, and all column sums are equal to one, i.e. $p_{ji} \geq 0$, and $\sum_{j=1}^{N} p_{ji} = 1$.

Based on the properties of the transition matrix $P$, one may differentiate between different types of Markov chains. A *non-homogenous* Markov chain has time-varying transition probabilities, i.e. the elements of $P$ will depend in some form on the time index $t$. If the transition matrix of a Markov chain can be partitioned in a block triangular form simply by rearranging its rows and columns it is called *reducible*, otherwise it is *irreducible*. Reducibility implies that one (or more) of the states do not *communicate* with each other in the sense that once the process enters state $j$ after being in state $i$ there is no possibility of returning to state $i$ at some future date.

Define the random variables $\xi_t^{(j)} = 1$ if $r_t = j$ and zero otherwise. Collecting the $N$ different values of $\xi_t^{(j)}$ at time $t$ in a column vector $\xi_t = [\xi_t^{(1)}, \ldots, \xi_t^{(N)}]'$, we may express any homogenous Markov chain as a first order vector autoregression (VAR) for $\xi_t$

---

[13] This Section draws heavily on the exposition given in Hamilton (1994), Chap. 22.2. For an exhaustive review of the statistical theory of Markov chains consider e.g. Karlin and Taylor (1975), or Ross (1996).

$$\xi_{t+1} = P \cdot \xi_t + v_{t+1}, \tag{A.29}$$

where the innovation process $v_{t+1}$ is a martingale difference sequence with expectation equal to zero. Thus, the $m$-step ahead forecast $\xi_{t+m}$, given $\xi_t$ is equal to

$$E(\xi_{t+m} \mid \xi_t) = P^m \cdot \xi_t, \tag{A.30}$$

where $P^m$ is a shorthand for multiplying the transition matrix $P$ by itself $m$ times. Note, that since all columns of the transition matrix sum to one, i.e. $P' \iota_N = \iota_N$ where $\iota_N$ denotes a $(N \times 1)$ vector of ones, $P$ has always a unit eigenvalue and $\iota_N$ is the associated eigenvector. A Markov chain is called *ergodic*, if $P$ has only one eigenvalue equal to unity and all other eigenvalues lie inside the unit circle. Ergodicity implies, that $P$ is associated with a column vector of ergodic probabilities $\pi = [\pi_1, \ldots, \pi_N]'$, and that $P^m$ converges to $\pi \cdot \iota_N'$, as $m \to \infty$. Thus, in combination with (A.30) we can deduce, that $\pi$ can be interpreted as the long run forecast or unconditional expectation of $\xi_{t+m}$, i.e.

$$E(\xi_{t+m}) = \pi.$$

For an ergodic Markov chain $\pi$ is unique and can be calculated from

$$\pi = (A'A)^{-1} A' \cdot e_{N+1}, \tag{A.31}$$

where the $(N + 1 \times N)$ matrix $A$ is given by

$$A = \begin{bmatrix} I_N - P \\ \iota_N' \end{bmatrix},$$

$I_N$ is the identity matrix of dimension $N$, and $e_{N+1}$ is the $(N + 1)$-th column of $I_{N+1}$.

An ergodic Markov chain is a covariance stationary process, despite the unit root of the transition matrix $P$, since the effects of past innovations $v_t, v_{t-1}, \ldots$ on $\xi_{t+m}$ can be shown to die out eventually, as $m \to \infty$. Furthermore, it can be shown that for an irreducible Markov chain all eigenvalues of $P$ lie either on or inside the unit circle, and that only one of the $N$ eigenvalues equals one. However, there might be more than one eigenvalue on the unit circle, implying that not all irreducible Markov chains are ergodic as well. If there are $k > 1$ eigenvalues on the unit circle, the Markov chain is called *periodic* with period $k$. Periodic Markov chains have the property that their states can be classified into $k$ distinct classes, so that if the process enters class $j$ at time $t$, it will leave this class at time $t + 1$ with certainty and return to class $j$ only at dates that are integer multiples of the period, i.e. $t + k, t + 2k, \ldots$.

Another important quantity that can be derived from the elements of the transition matrix $P$ is the *expected duration* of staying in a given state $j$, once the process enters it. Assume that the process enters state $j$ at time $t$. Then the duration $D_j$ will be equal to one, if $r_{t+1} \neq j$, which will happen with probability $p(D_j = 1) = (1 - p_{jj})$. In general, the event $D_j = m$ will be observed, if $r_{t+1} = j, ..., r_{t+m-1} = j$ and $r_{t+m} \neq j$, so $p(D_j = m) = p_{jj}^{m-1} \cdot (1 - p_{jj})$. Thus, the expected duration of state $j$ is given by[14]

$$E(D_j) = \sum_{m=1}^{\infty} m \cdot p_{jj}^{m-1} \cdot (1 - p_{jj}) \qquad (A.32)$$

$$= \frac{1}{1 - p_{jj}},$$

if $p_{jj} < 1$ and $E(D_j) = \infty$, if $p_{jj} = 1$.

Maximum likelihood estimates of the transition probabilities $p_{ji}$ can be derived as follows. Consider observing a time series of length $T$ on the discrete valued process $r_t$. In a first step we compute the indicator functions $\xi_t^{(j)}$ and $\xi_t^{(ji)} \equiv \xi_t^{(j)} \cdot \xi_{t-1}^{(i)}$. Note, that $\xi_t^{(ji)}$ is equal to one, if state $i$ was followed by state $j$ and zero otherwise. If the Markov chain is homogenous, the probability of observing a transition from state $i$ to state $j$ will be equal to $p_{ji}$ for all $t$. Thus, conditioning on the initial state at time $t = 1$, the likelihood function for a sample of $T$ observations may be written as

$$\mathcal{L}(\theta) = \prod_{t=2}^{T} \prod_{j=1}^{N} \prod_{i=1}^{N} p_{ji}^{\xi_t^{(ji)}},$$

with parameter vector $\theta = (p_{11}, \ldots, p_{NN})$. The associated log likelihood function is

$$\ln \mathcal{L}(\theta) = \sum_{t=2}^{T} \sum_{j=1}^{N} \sum_{i=1}^{N} \xi_t^{(ji)} \cdot \ln(p_{ji}).$$

Estimates of the transition probabilities may be obtained by maximizing $\ln \mathcal{L}(\theta)$ subject to $N$ linear constraints of the form

$$\sum_{j=1}^{N} p_{ji} = 1.$$

This leads to the following first order conditions

$$\frac{1}{p_{ji}} \cdot \sum_{t=2}^{T} \xi_t^{(ji)} - \gamma_i \stackrel{!}{=} 0,$$

where $\gamma_i$ is the Lagrange multiplier for the $i$-th constraint. Rearranging and summing over all $N$ possible states we find that

$$\gamma_i = \sum_{t=2}^{T} \sum_{j=1}^{N} \xi_t^{(ji)} = \sum_{t=2}^{T} \xi_{t-1}^{(i)},$$

---

[14] See Kim and Nelson (1999), pp. 71-72.

so the corresponding maximum likelihood estimator is given by

$$\hat{p}_{ji} = \frac{\sum_{t=2}^{T} \xi_t^{(ji)}}{\sum_{t=2}^{T} \xi_{t-1}^{(i)}}. \tag{A.33}$$

It can be shown, that this estimator coincides with the ordinary least squares estimator of $p_{ji}$ for the VAR representation (A.29), see Gouriéroux and Jasiak (2001), p. 235.

## A.9 The Smoothing Algorithm

The recursive algorithm employed for the calculation of the smoothed inferences on the state of the regime given the full sample of observable quantities $\hat{\xi}_{t|T}^{(j)} \equiv p(r_t = j \mid \mathcal{Y}_T; \theta)$ has been proposed by Kim (1994) in the context of an autoregressive state-space model with time-varying regression coefficients whose data generating process is governed by an unobservable Markov chain. The calculations needed to derive the smoothing algorithm are analogous to the Kalman filter algorithm. The following derivation is based on the exposition of Kim's algorithm given in Hamilton (1994), pp. 700-702.

Let us define the information set $\mathcal{Y}_t = (y_t, y_{t-1}, \ldots, y_1, x_t, x_{t-1}, \ldots, x_1)$. Note, that the backward recursion of (4.36) starts with $\hat{\xi}_{T|T}^{(j)}$, the filtered inference on the state of the regime at the end of the sample obtained from (4.33) and (4.34). The first step in deriving (4.36) is to notice that[15]

$$
\begin{aligned}
p(r_t = j \mid r_{t+1} = i; \mathcal{Y}_{t+1}) &= \frac{p(r_t = j \cap y_{t+1} \mid r_{t+1} = i; x_{t+1}, \mathcal{Y}_t)}{f(y_{t+1} \mid r_{t+1} = i; x_{t+1}, \mathcal{Y}_t)} \\
&= p(r_t = j \mid r_{t+1} = i; x_{t+1}, \mathcal{Y}_t) \\
&\quad \cdot \frac{f(y_{t+1} \mid r_t = j, r_{t+1} = i; x_{t+1}, \mathcal{Y}_t)}{f(y_{t+1} \mid r_{t+1} = i; x_{t+1}, \mathcal{Y}_t)} \\
&= p(r_t = j \mid r_{t+1} = i; x_{t+1}, \mathcal{Y}_t) \\
&= p(r_t = j \mid r_{t+1} = i; \mathcal{Y}_t),
\end{aligned}
$$

since $y_{t+1}$ depends on the sequence of regime variables $r_{t+1}, r_t, r_{t-1} \ldots$ only through its current realization $r_{t+1}$, and $r_t$ is independent of $x_{t+1}$. Similar calculations show that

$$p(r_t = j \mid r_{t+1} = i; \mathcal{Y}_{t+m}) = p(r_t = j \mid r_{t+1} = i; \mathcal{Y}_t) \tag{A.34}$$

holds for $m = 2, 3 \ldots$ as well. Furthermore we know that

---

[15] In the following derivation, the dependence on the parameter vector $\theta$ has been suppressed for notational simplification.

$$p(r_t = j \mid r_{t+1} = i; \mathcal{Y}_{t+m}) = \frac{p(r_t = j \cap r_{t+1} = i \mid \mathcal{Y}_t)}{p(r_{t+1} = i \mid \mathcal{Y}_t)} \tag{A.35}$$

$$= \frac{p(r_t = j \mid \mathcal{Y}_t) \cdot p(r_{t+1} = i \mid r_t = j)}{p(r_{t+1} = i \mid \mathcal{Y}_t)}$$

$$= \frac{p(r_t = j \mid \mathcal{Y}_t) \cdot p_{ji}}{p(r_{t+1} = i \mid \mathcal{Y}_t)}.$$

Now, combining these two results, we can show that

$$p(r_t = j \cap r_{t+1} = i \mid \mathcal{Y}_T) = p(r_{t+1} = i \mid \mathcal{Y}_T) \cdot p(r_t = j \mid r_{t+1} = i; \mathcal{Y}_T)$$

$$= p(r_{t+1} = i \mid \mathcal{Y}_T) \cdot p(r_t = j \mid r_{t+1} = i; \mathcal{Y}_t)$$

$$= p(r_{t+1} = i \mid \mathcal{Y}_T) \cdot \frac{p_{ji} \cdot p(r_t = j \mid \mathcal{Y}_t)}{p(r_{t+1} = i \mid \mathcal{Y}_t)},$$

or, employing our previous notation

$$\hat{\xi}_{t+1|T}^{(ij)} = \hat{\xi}_{t+1|T}^{(i)} \cdot \frac{p_{ij} \cdot \hat{\xi}_{t|t}^{(j)}}{\hat{\xi}_{t+1|t}^{(i)}}. \tag{A.36}$$

which is equivalent to (4.38). Equation (4.36) may be calculated from this expression immediately by noting that

$$p(r_t = j \mid \mathcal{Y}_T) = \sum_{i=1}^{N} p(r_t = j \cap r_{t+1} = i \mid \mathcal{Y}_T).$$

## A.10 Estimation of Transition Probabilities in the Markov Switching Model

In order to find parameter estimates that satisfy the restriction, that the column sums of the Markov transition matrix $P$ are equal to one, we maximize the following Lagrange function $J(\theta)$

$$J(\theta) = \ln \mathcal{L}_{EC}(\theta) + \sum_{i=1}^{N} \gamma_i \cdot \left( 1 - \sum_{j=1}^{N} p_{ji} \right), \tag{A.37}$$

where $\ln \mathcal{L}_{EC}(\theta)$ is defined in (4.37). Differentiating (A.37) with respect to $p_{ji}$ yields

$$\frac{\partial J(\theta)}{\partial p_{ji}} = \frac{\partial \left( \sum_{t=2}^{T} \sum_{j=1}^{N} \sum_{i=1}^{N} \hat{\xi}_{t|T}^{(ji)} \cdot \ln p_{ji} \right)}{\partial p_{ji}} + \frac{\partial \left( \sum_{i=1}^{N} \gamma_i \cdot \left( 1 - \sum_{j=1}^{N} p_{ji} \right) \right)}{\partial p_{ji}}$$

$$= \frac{1}{p_{ji}} \cdot \sum_{t=2}^{T} \hat{\xi}_{t|T}^{(ji)} - \gamma_i \overset{!}{=} 0.$$

Rearranging and summing over all $N$ possible states leads to

$$\gamma_i = \sum_{t=2}^{T} \sum_{j=1}^{N} \hat{\xi}_{t|T}^{(ji)} = \sum_{t=2}^{T} \hat{\xi}_{t-1|T}^{(i)},$$

since $\sum_{j=1}^{N} p_{ji} = 1$, so that by replacing $\gamma_i$ we obtain

$$\hat{p}_{ji} = \frac{\sum_{t=2}^{T} \hat{\xi}_{t|T}^{(ji)}}{\sum_{t=2}^{T} \hat{\xi}_{t-1|T}^{(i)}} = \frac{\sum_{t=2}^{T} p\left(r_t = j \cap r_{t-1} = i \mid \mathcal{Y}_T; \theta\right)}{\sum_{t=2}^{T} p\left(r_{t-1} = i \mid \mathcal{Y}_T; \theta\right)}, \qquad \text{(A.38)}$$

which is the expression given in (4.39). The estimates of the initial regime probabilities $p_j$ may be obtained by forming the Lagrange function

$$J(\theta)^* = \ln \mathcal{L}_{EC}^*(\theta) + \gamma_0 \cdot \left(1 - \sum_{j=1}^{N} p_j\right) + \sum_{i=1}^{N} \gamma_i \cdot \left(1 - \sum_{j=1}^{N} p_{ji}\right), \qquad \text{(A.39)}$$

and setting its derivative equal to zero. From

$$\frac{\partial J(\theta)^*}{\partial p_j} = \frac{1}{p_j} \cdot \hat{\xi}_{1|T}^{(j)} - \gamma_0 \overset{!}{=} 0,$$

we find by summing over the $N$ possible states of the regime variable that $\gamma_0 = 1$, so that

$$\hat{p}_j = \hat{\xi}_{1|T}^{(j)}. \qquad \text{(A.40)}$$

## A.11 Moments of the Dependent Variable in a Markov Switching Model

Noting, that the density of the observable variable $y_t$ in the Markov switching model is given by[16]

$$f(y_t) = \sum_{j=1}^{N} \xi_{t|t-1}^{(j)} \cdot f_j(y_t \mid r_t = j; \theta),$$

we may immediately deduce, that all of the expressions given in Appendix A.6 for the moments of the static mixture model are valid for the Markov switching

---

[16] See (4.35).

model as well, after replacing the time independent regime probabilities $\pi_j$ by the time varying probabilities $\xi_{t|t-1}^{(j)}$. In particular, note that the conditional mean and the conditional variance of a Markov switching regression model are given by

$$E(y_t \mid x_t; \theta) = \sum_{j=1}^{N} \xi_{t|t-1}^{(j)} \cdot \lambda_{tj} \cdot \Delta t$$

$$Var(y_t \mid x_t; \theta) = \sum_{j=1}^{N} \xi_{t|t-1}^{(j)} \cdot Var(y_t \mid r_t = j, x_t; \theta)$$

$$+ \sum_{j=1}^{N} \xi_{t|t-1}^{(j)} \cdot [\lambda_{tj} \cdot \Delta t - E(y_t \mid x_t; \theta)]^2,$$

with regime specific conditional variance function $Var(y_t \mid r_t = j, x_t; \theta) = \lambda_{tj} \cdot \Delta t$, in the Poisson model, and $Var(y_t \mid r_t = j, x_t; \theta) = \lambda_{tj} \cdot \Delta t + \alpha_j \cdot (\lambda_{tj} \cdot \Delta t)^2$, when a Negbin 2 model has been specified.

An important distinction between the Markov switching and the static framework is the serial dependence structure. The static mixture model assumes[17], that autocovariances for all lags are identically equal to zero. In the Markov switching model however it can be shown that autocorrelation between successive realizations of the observed variable arises solely as a consequence of the dependence of $y_t$ on the latent regime $r_t$, which itself follows an autoregressive process. The derivation of the conditional autocovariance function for a Markov switching regression model is quite cumbersome.[18] For $m > 0$ it is given by

$$Cov(y_t, y_{t+m} \mid x_t; \theta) = \sum_{j=1}^{N} \sum_{i=1}^{N} \pi_i \cdot p_{ji}^m \cdot \lambda_{ti} \cdot \lambda_{(t+m)j} \cdot (\Delta t)^2$$

$$- \left( \sum_{j=1}^{N} \pi_j \cdot \lambda_{tj} \cdot \Delta t \right) \cdot \left( \sum_{j=1}^{N} \pi_j \cdot \lambda_{(t+m)j} \cdot \Delta t \right),$$

where $\pi_j$ denotes the ergodic probability of being in state $j$, and $p_{ji}^m$ denotes the row $j$, column $i$ element of the matrix $P^m$, which is our shorthand for multiplying the transition matrix $P$ by itself $m$ times, see Appendix A.8. Note, that if no covariates $x_t$ have been specified, so that the intensities $\lambda_j$ are time invariant, the autocovariances (and accordingly, the autocorrelation function) depend on the lag length $m$ only through the elements of the matrix $P^m$.

---

[17] We assume that no autoregressive dynamics are being specified via the inclusion of lagged dependent variables in the conditional mean specification.
[18] See MacDonald and Zucchini (1997), pp. 69-77, and pp. 128-129, and Timmermann (2000) for details.

# References

Abramowitz, M., and I.A. Stegun, 1965, *Handbook of mathematical functions.* (Dover Publications Inc. New York).

Admati, A. R., and P. Pfleiderer, 1988, A Theory of intraday patterns: Volume and price variability, *Review of Financial Studies* 1, 3–40.

Admati, A. R., and P. Pfleiderer, 1989, Divide and conquer: A Theory of intraday and day-of-the-week mean effects, *Review of Financial Studies* 2, 189–223.

Aitken, M., and A. Frino, 1996, The accuracy of the tick test: Evidence from Australian stock exchange, *Journal of Banking & Finance* 20, 1715–1729.

Aitken, M., and D. B. Rubin, 1985, Estimation and hypothesis testing in finite mixture models, *Journal of the Royal Statistical Society, Series B* 47, 67–75.

Akaike, H., 1973, Information theory and an extension of the maximum likelihood principle, in N. Petrov, and F. Csàdki, eds.: *Proceedings of the 2nd International Symposium on Information Theory* (Akadémiai Kiadó, Budapest ).

Aktas, N., E. de Bodt, F. Declerck, and H. Van Oppens, 2003, Probability of informed trading? Some evidences around corporate events, Working Paper Université Catholique de Louvain.

Allenby, G. M., R. P. Leone, and L. Jen, 1999, A dynamic model of purchase timing with application to direct marketing, *Journal of the American Statistical Association* 94, 365–374.

Andersen, T. G., and T. Bollerslev, 1997, Intraday periodicity and volatility persistence in financial markets, *Journal of Empirical Finance* 4, 115–158.

Anderson, T. W., and A. M. Walker, 1964, On the asymptotic distribution of a sample from a linear stochstic process, *Annals of Mathematical Statistics* 35, 1269–1303.

Andrews, D. W. K., 1988a, Chi-square diagnostic tests for econometric models: Introduction and applications, *Journal of Econometrics* 37, 135–156.

Andrews, D. W. K., 1988b, Chi-square diagnostic tests for econometric models: Theory, *Econometrica* 56, 1419–1453.

Andrews, D. W. K., 1991, Asymptotic normality of series estimators for non-parametric and semiparametric regression models, *Econometrica* 59, 307–345.

Andrews, D. W. K., and W. Ploberger, 1994, Optimal tests when a nuisance parameter is present only under the alternative, *Econometrica* 62, 1383–1414.

Ang, A., and G. Bekaert, 2002, Regime switches in interest rates, *Journal of Business & Economic Statistics* 20, 163–182.

Bagehot, W., 1971, The only game in town, *Financial Analysts Journal* 27, 12–14, 22.

Bauwens, L., and P. Giot, 2001, *Econometric modelling of stock market intraday activity*. (Kluwer Acadenic Publishers Boston).

Beber, A., and C. Caglio, 2002, Order submission strategies and information: Empirical evidence from the NYSE, Working Paper Università Bocconi Milano.

Bekaert, G., and C. R. Harvey, 1995, Time-varying world market integration, *The Journal of Finance* 50, 403–444.

Bickel, P. J., and Y. Ritov, 1996, Inference in hidden Markov models I: Local asymptotic normality in the stationary case, *Bernoulli* 2, 199–228.

Blomberg, S. B., 2000, Modeling political change with a regime-switching model, *European Journal of Political Economy* 16, 739–762.

Böhning, D., 1995, A review of reliable maximum likelihood algorithms for semiparametric mixture models, *Journal of Statistical Planning and Inference* 47, 5–28.

Böhning, D., E. Dietz, R. Schaub, P. Schlattman, and B. G. Lindsay, 1994, The distribution of the likelihood ratio for mixtures of densities from the one-parameter exponential family, *Annals of the Institute of Statistical Mathematics* 46, 373–388.

Bollerslev, T., 1986, Generalized autoregressive conditional heteroscedasticity, *Journal of Econometrics* 31, 307–327.

Boswell, M. T., and G. P. Patil, 1970, Chance mechanisms generating the negative binomial distributions, in G. P. Patil, eds.: *Random Counts in Scientific Work, Vol. 1: Random Counts in Models and Structures* (Pennsylvenia University Press, University Park ).

Box, G. E. P., and D. A. Pierce, 1970, Distribution of residual autocorrelations in autoregressive-integrated moving average times series models, *Journal of the American Statistical Association* 65, 1509–1526.

Boyles, R. A., 1983, On the convergence of the EM algorithm, *Journal of the Royal Statistical Society, Series B* 45, 47–50.

Brock, W. A., and A. W. Kleidon, 1992, Periodic market closure and trading volume, *Journal of Economic Dynamics and Control* 16, 451–489.

Brown, P., N. Thomson, and D. Walsh, 1999, Characteristics of the order flow through an electronic open limit order book, *Journal of International Financial Markets, Institutions and Money* 9, 335–357.

Cai, J., 1994, A Markov model of switching-regime ARCH, *Journal of Business & Economic Statistics* 12, 309–316.

Cameron, A. C., and P. K. Trivedi, 1986, Econometric models based on count data: Comparisons and applications of some estimators and tests, *Journal of Applied Econometrics* 1, 29–53.

Cameron, A. C., and P. K. Trivedi, 1998, *Regression analysis of count data.* (Cambridge University Press Cambridge).

Chauvet, M., C. Juhn, and S. Potter, 2002, Markov switching in disaggregate unemployment rates, *Empirical Economics* 27, 205–232.

Chordia, T., R. Roll, and A. Subrahmanyam, 2002, Order imbalance, liquidity, and market returns, *Journal of Financial Economics* 65, 111–130.

Chow, G. C., 1960, Tests of equality between sets of coefficients in two linear regressions, *Econometrica* 28, 591–605.

Chung, K. H., B. F. Van Ness, and R. A. Van Ness, 1999, Limit orders and the bid-ask spread, *Journal of Financial Economics* 53, 255–287.

Coe, P. J., 2002, Power issues when testing the Markov switching model with the sup likelihood ratio test using U.S. output, *Empirical Economics* 27, 395–401.

Cohen, K. J., S. F. Maier, R. A. Schwartz, and D. K. Whitcomb, 1986, *The Microstructure of Securities Markets.* (Prentice Hall Englewodd Cliffs).

Copeland, T. E., and D. Galai, 1983, Information effects on the bid-ask spread, *Journal of Finance* 38, 1457–1469.

Cosslett, S. R., and L.-F. Lee, 1985, Serial correlation in latent discrete variable models, *Journal of Econometrics* 27, 79–97.

Coughenour, J., and K. Shastri, 1999, Symposium on market microstructure: A review of empirical research, *Financial Review* 34, 1–28.

Davidson, R., and J. G. MacKinnon, 1993, *Estimation and inference in econometrics.* (Oxford University Press New York).

Davies, R. B., 1977, Hypothesis testing when a nuisance parameter is present only under the alternative, *Biometrika* 64, 247–254.

Davies, R. B., 1987, Hypothesis testing when a nuisance parameter is present only under the alternatives, *Biometrika* 74, 33–43.

De Jong, C., 2001, The impact of option trading on the dynamics of the underlying price process, Working Paper Erasmus University Rotterdam.

Deb, P., X. Ming, and P. K. Trivedi, 1999, A comparison of maximum likelihood and moment-based estimators of finite mixture count models, Discussion Paper Department of Economics, Indiana University.

Deb, P., and P. K. Trivedi, 2002, The structure of demand for health care: Latent class versus two-part models, Working Paper Indiana University-Purdue University Indianapolis.

Dempster, A. P., N. M. Laird, and D. B. Rubin, 1977, Maximum likelihood from incomplete data via the EM algorithm (with discussion), *Journal of the Royal Statistical Society, Series B* 39, 1–38.

Dewachter, H., 2001, Can Markov switching models replicate chartist profits in the foreign exchange market?, *Journal of International Money and Finance* 20, 25–41.

Diamond, D. W., and R. E. Verrecchia, 1987, Constraints on short-selling and asset price adjustments to private information, *Journal of Financial Economics* 18, 227–311.

Diebold, F. X., J.-H. Lee, and G. C. Weinbach, 1997, Regime switching with time-varying transition probabilities, in C. P. Hargreaves, eds.: *Nonstationary Time Series Analysis and Cointegration* (Oxford University Press, Oxford ).

Diebold, F. X., and G. D. Rudebusch, 1996, Measuring business cycles: A modern perspective, *Review of Economics and Statistics* 78, 67–77.

Diebolt, J., and C. P. Robert, 1994, Estimation of finite mixture distributions through Bayesian sampling, *Journal of the Royal Statistical Society, Series B* 56, 363–375.

Domowitz, I., 1993, A taxonomy of automated trade execcution systems, *Journal of International Money and Finance* 12, 607–631.

Douc, R., and C. Matias, 2001, Asymptotics of the maximum likelihood estimator for general hidden Markov models, *Bernoulli* 7, 381–420.

D'Unger, A. V., K. C. Land, and P. L. McCall, 2002, Sex differences in age patterns of delinquent/criminal careers: Results from Poisson latent class analyses of the Philadelphia cohort study, *Journal of Quantitative Criminology* 18, 349–375.

D'Unger, A. V., K. C. Land, P. L. McCall, and D. S. Nagin, 1998, How many latent classes of delinquent/criminal careers? Results from mixed Poisson regression analyses, *American Journal of Sociology* 103, 1593–1630.

Easley, D., S. Hvidkjaer, and M. O'Hara, 2001, Is information risk a determinant of asset returns?, Working Paper Cornell University.

Easley, D., N. Kiefer, M. O'Hara, and J. P. Paperman, 1996, Liquidity, information and infrequently traded stocks, *Journal of Finance* 51, 1405–1436.

Easley, D., N. M. Kiefer, and M. O'Hara, 1996, Cream-skimming or profit-sharing? The curious role of puchased order flow, *Journal of Finance* 51, 811–833.

Easley, D., N. M. Kiefer, and M. O'Hara, 1997a, The information content of the trading process, *Journal of Empirical Finance* 4, 159–186.

Easley, D., N. M. Kiefer, and M. O'Hara, 1997b, One day in the life of a very common stock, *The Review of Financial Studies* 10, 805–835.

Easley, D., and M. O'Hara, 1987, Price, trade size, and information in securities markets, *Journal of Financial Economics* 19, 69–90.

Easley, D., and M. O'Hara, 1991, Order form and information in securities markets, *Journal of Finance* 46, 905–927.

Easley, D., and M. O'Hara, 1992a, Adverse selection and large trade volume: The implications for market efficiency, *Journal of Financial and Quantitaive Analysis* 27, 185–208.

Easley, D., and M. O'Hara, 1992b, Time and the process of security price adjustment, *Journal of Finance* 47, 577–605.

Easley, D., and M. O'Hara, 1995, Market microstructure, in R. A. Jarrow, V. Maksimovic, and W. T. Ziemba, eds.: *Handbooks in Operations Research and Managment Science, Vol. 9: Finance* (Elsevier, Amsterdam ).

Easley, D., M. O'Hara, and J. Paperman, 1998, Financial analysts and information-based trade, *Journal of Financial Markets* 1, 1751–201.

Easley, D., M. O'Hara, and G. Saar, 2001, How stock splits affect trading: A microsctructure approach, *Journal of Financial and Quantitaive Analysis* 36, 25–51.

Easley, D., M. O'Hara, and P. S. Srinivas, 1998, Option volume and stock prices: Evidence on where informed traders trade, *Journal of Finance* 53, 431–465.

Easley, D. E., R. F. Engle, M. O'Hara, and L. Wu, 2002, Time-varying arrival rates of informed and uninformed trades, Working Paper Fordham University.

Eastwood, B. J., 1991, Asymptotic normality and consistency of seminonparametric regression estimators using an upwards F test truncation rule, *Journal of Econometrics* 20, 151–181.

Ellis, K., R. Michaely, and M. O'Hara, 2000, The accuracy of trade classification rules: Evidence from Nasdaq, *Journal of Financial and Quantitative Analysis* 35, 529–551.

Engel, C., and J. D. Hamilton, 1990, Long swings in the dollar: Are they in the data and do markets know it?, *American Economic Review* 80, 689–713.

Engle, R. F., and J. R. Russell, 1998, Autoregressive conditional duration: A new model for irregulary spaced transaction data, *Econometrica* 66, 1127–1162.

Eubank, R. L., and P. Speckman, 1990, Curve fitting by polynomial-trigonometric regression, *Biometrika* 77, 1–9.

Feng, Z. D., and C. E. McCulloch, 1996, Using bootstrap likelihood ratios in finite mixture models, *Journal of the Royal Statistical Society, Series B* 58, 609–617.

Filardo, A. J., 1994, Business-cycle phases and their transitional dynamics, *Journal of Business and Economic Statistics* 12, 299–308.

Finucane, T. J., 2000, A direct test of methods for inferring trade direction from intra-day data, *Journal of Financial and Quantitative Analysis* 35, 553–576.

Forster, M. M., and T. J. George, 1992, Anonymity in securities markets, *Journal of Financial Intermediation* 2, 168–206.

Foster, F. D., and S. Viswanathan, 1990, A theory of the intraday variations in volume, variance, and trading costs in securities markets, *Review of Financial Studies* 3, 593–624.

Foster, F. D., and S. Viswanathan, 1993, The effect of public information and competition on trading volume and price volatility, *Review of Financial Studies* 6, 23–56.

Fu, H., 2002, Information asymmetry and the market reaction to equity carve-outs, Working Paper University of Minnesota.

Gallant, A. R., 1981, On the bias in flexible functional forms and an essentially unbiased form, *Journal of Econometrics* 20, 285–323.

Garcia, R., 1998, Asymptotic null distribution of the likelihood ratio test in Markov switching models, *International Economic Review* 39, 763–788.

Garcia, R., and P. Perron, 1996, An analysis of the real interest rate under regime shifts, *The Review of Economics and Statistics* 78, 111–125.

Glosten, L., and P. Milgrom, 1985, Bid, ask, and transaction prices in a specialist market with heterogeneously informed traders, *Journal of Financial Economics* 14, 71–100.

Glosten, L. R., 1994, Is the electronic limit order book inevitable?, *Journal of Finance* 49, 1127–1161.

Goldfeldt, S. M., and R. E. Quandt, 1965, Some tests for homocscedasticity, *Journal of the American Statistical Association* 60, 539–547.

Goldfeldt, S. M., and R. E. Quandt, 1973, A Markov model for switching regressions, *Journal of Econometrics* 1, 3–16.

Goodhart, C. A. E., and M. O'Hara, 1997, High frequency data in financial markets: Issues and applications, *Journal of Empirical Finance* 4, 73–114.

Gouriéroux, C., and J. Jasiak, 2001, *Financial Econometrics*. (Princeton University Press Princeton).

Grammig, J., R. Hujer, and S. Kokot, 2002, Tackling boundary effects in nonparametric estimation of intra-day liquidity measures, *Computational Statistics* 17, 233–249.

Grammig, J., D. Schiereck, and E. Theissen, 1999, Informationsbasierter Aktienhandel über IBIS, Working Paper Series: Finance & Accounting University of Frankfurt.

Grammig, J., D. Schiereck, and E. Theissen, 2001, Knowing me, knowing you: Trader anonymity and informed trading in parallel markets, *Journal of Financial Markets* 4, 385–412.

Grammig, J., and E. Theissen, 2002, Estimating the probability of informed trading - Does trade misclassification matter?, Working Paper University of St. Gallen.

Gray, S., 1996, Modelling the conditional distribution of interest rates as a regime-switching process, *Journal of Financial Economics* 42, 27–62.

Greene, W. H., 1994, Accounting for excess zeros and sample selection in Poisson and negative binomial regression models, Working Paper No. EC-94-10 New York University.

Gritz, R. M., 1993, The impact of training on the frequency and duration of employment, *Journal of Econometrics* 57, 21–51.

Grossman, S. J., and J. E. Stiglitz, 1980, On the impossibility of informationally efficient markets, *American Economic Review* 70, 393–408.

Hamilton, J. D., 1989, A new approach to the economic analysis of nonstationary time series and the business cycle, *Econometrica* 57, 357–384.

Hamilton, J. D., 1990, Analysis of time series subject to changes in regime, *Journal of Econometrics* 45, 39–70.

Hamilton, J. D., 1991, A quasi-Bayesian approach to estimating parameters for mixtures of normal distributions, *Journal of Business and Economic Statistics* 9, 27–39.

Hamilton, J. D., 1993, Estimation, inference and forecasting of time series subject to changes in regime, in G. S. Maddala, C. R. Rao, and H. D. Vinod, eds.: *Handbook of Statistics Vol. 11* (Elsevier, Amsterdam ).

Hamilton, J. D., 1994, *Time series analysis.* (Princeton University Press Princeton).

Hamilton, J. D., 1996, Specification testing in Markov-switching time-series models, *Journal of Econometrics* 70, 127–157.

Hamilton, J. D., and G. Perez-Quiros, 1996, What do the leading indicators lead?, *Journal of Business* 69, 27–49.

Hamilton, J. D., and B. Raj, 2002, New directions in business cycle research and financial analysis, *Empirical Economics* 27, 149–162.

Hamilton, J. D., and R. Susmel, 1994, Autoregressive conditional heteroscedasticity and changes in regime, *Journal of Econometrics* 64, 307–333.

Handa, P., and R. A. Schwartz, 1996, Limit order trading, *Journal of Finance* 51, 1835–1861.

Hansen, B. E., 1992, The likelihood ratio test under nonstandard conditions: Testing the Markov switching model of GNP, *Journal of Applied Econometrics* 7, S61–S82.

Hansen, B. E., 1996a, Erratum: The likelihood ratio test under nonstandard conditions: Testing the Markov switching model of GNP, *Journal of Applied Econometrics* 11, 195–198.

Hansen, B. E., 1996b, Inference when a nuisance parameter is not identified under the Null hypothesis, *Econometrica* 64, 413–430.

Hasbrouck, J., 1992, Using the TORQ database, Working Paper, New York University, Stern School of Business.

Hasbrouck, J., 1996, Modeling market microstructure time series, in G. S. Maddala, and C. R. Rao, eds.: *Handbooks of Statistics, Vol. 14: Statistical Methods in Finance* (Elsevier, Amsterdam ).

Hasbrouck, J., 1999, Trading fast and slow: Security market events in real time, Working Paper New York University.

Hasbrouck, J., G. Sofianos, and D. Sosebee, 1993, New York Stock Exchange systems and trading procedures, Working Paper # 93-01 New York Stock Exchange.

Hasbrouk, J., and G. Sofianos, 1993, The trades of market makers: An empirical analysis of NYSE specialists, *The Journal of Finance* 48, 1565–1593.

Hausman, J., A. Lo, and C. MacKinlay, 1992, An ordered probit analysis of transaction stock prices, *Journal of Financial Economics* 31, 319–379.

Heckman, J. J., 1984, The $\chi^2$ goodness of fit test for models with parameters estimated from microdata, *Econometrica* 52, 1543–1548.

Heckman, J. J., R. Robb, and J. R. Walker, 1990, Testing the mixture of exponentials hypothesis and estimating the mixing distribution by the method of moments, *Journal of the American Statistical Association* 85, 582–589.

Heckman, J. J., and B. Singer, 1982, A method for minimizing the impact of distributional assumptions in econometric models for duration data, *Econometrica* 52, 271–320.

Heckman, J. J., and C. R. Taber, 1994, Econometric mixture models and more general models for unobservables in duration analysis, NBER Technical Working Paper No. 157 National Bureau of Economic Research.

Heidle, H. G., and R. D. Huang, 1999, Information-based trading in dealer and auction markets: An analysis of exchange listings, Working Paper Owen Graduate School of Management, Vanderbilt University.

Holden, C. W., and A. Subrahmanyam, 1992, Long-lived private information and imperfect competition, *Journal of Finance* 47, 247–270.

Holden, R. T., 1987, Time series analysis of a contagious process, *Journal of the American Statistical Association* 82, 1019–1026.

Hujer, R., J. Grammig, and S. Kokot, 2000, Time varying trade intensities and the Deutsche Telekom IPO, *Journal of Economics and Statistics* 220, 689–714.

Hujer, R., S. Vuletić, and S. Kokot, 2002, The Markov switching ACD model, Working Paper Series: Finance & Accounting No. 90 University of Frankfurt.

Jern, B., 2000, Causes of observed feedback patterns between stocks and options, Working Paper Swedish School of Economics and Business Administration.

Jewell, N. P., 1982, Mixtures of exponential distributions, *The Annals of Statistics* 10, 479–484.

John, K., A. Koticha, R. Narayanan, and M. Subrahmanyam, 2000, Margin rules, informed trading in derivatives, and price dynamics, Working Paper New York University.

Johnson, N. L., and S. Kotz, 1969, *Discrete Distributions*. (John Wiley New York).

Karlin, S., and H. M. Taylor, 1975, *A first course in stochastic processes*. (Academic Press Cambridge) 2nd edn.

Kim, C.-J., 1994, Dynamic linear models with Markov-switching, *Journal of Econometrics* 60, 1–22.

Kim, C.-J., and C. R. Nelson, 1999, *State-space models with regime switching*. (MIT Press Cambridge).

Kingman, J. F. C., 1993, *Poisson Processes*. (Oxford University Press Oxford).

Krolzig, H.-M., M. Marcellino, and G. E. Mizon, 2002, A Markov-switching vector equilibrium correction model of the UK labour market, *Empirical Economics* 27, 233–254.

Kyle, A. S., 1984, Market structure, information, futures markets, and price formation, in G. G. Story, A. Schmitz, and A. H. Sarris, eds.: *International agricultural trade: Advanced readings in price formation, market structure, and price instability* (Westview Press, Boulder and London ).

Kyle, A. S., 1985, Continuous auctions and insider trading, *Econometrica* 53, 1315-1336.

Kyle, A. S., 1989, Informed speculation with imperfect competition, *Review of Economic Studies* 56, 317-356.

Laird, N., 1978, Nonparametric maximum likelihood estimation of a mixing distribution, *Journal of the American Statistical Association* 73, 805-811.

Lam, P.-S., 1990, The Hamilton model with a general autoregressive component, *Journal of Monetary Economics* 26, 409-432.

Lambert, D., 1992, Zero-inflated Poisson regresion, with an application to defects in manufactoring, *Technometrics* 34, 1-14.

Land, K. C., D. S. Nagin, and P. L. McCall, 2001, Discrete-time hazard regression models with hidden heterogeneity: The semiparametric mixed Poisson regression approach, *Sociological Methods & Research* 29, 342-373.

Lawless, J. F., 1987, Negative binomial and Poisson regression, *The Canadian Journal of Statistics* 15, 209-225.

Lee, C. M. C., and B. Radhakrishna, 2000, Inferring investor behaviour: Evidence from TORQ data, *Journal of Financial Markets* 3, 83-111.

Lee, C. M. C., and M. J. Ready, 1991, Inferring trade direction from intraday data, *The Journal of Finance* 46, 733-746.

Lei, Q., and G. Wu, 2001, The behavior of uninformed investors and time-varying informed trading activities, Working Paper Fordham University.

Leroux, B. G., 1992, Consistent estimation of a mixing distribution, *The Annals of Statistics* 20, 1350-1360.

Li, W. K., 1994, Time series models based on generalized linear models: Some further results, *Biometrics* 50, 506-511.

Lindgren, B. W., G. W. McElbrath, and D. A. Berry, 1978, *Introduction to probability and statistics.* (Macmillan New York) 4th edn.

Lindgren, G., 1978, Markov regime switching models for mixed distributions and switching regressions, *Scandinavian Journal of Statistics* 5, 81-91.

Lindsay, B. G., and M. L. Lesperance, 1995, A review of semiparametric mixture models, *Journal of Statistical Planning and Inference* 47, 29-39.

Lindsay, B. G., and K. Roeder, 1992, Residual diagnostics for mixture models, *Journal of the American Statistical Association* 87, 785-794.

Ljung, G. M., and G. E. P. Box, 1978, On a measure of lack of fit in time series models, *Biometrika* 65, 297-303.

MacDonald, I. L., and W. Zucchini, 1997, *Hidden markov and other models for discrete-valued time series.* (Chapman & Hall London).

Madhavan, A., 1992, Trading mechanisms in securities markets, *Journal of Finance* 47, 607-641.

Madhavan, A., 2000, Market microstructure: A survey, *Journal of Financial Markets* 3, 205-258.

McCulloch, R. E., and R. S. Tsay, 1994, Statistical analysis of economic time series via Markov switching models, *Journal of Time series Analysis* 15, 523–539.

Milgrom, P., and N. Stokey, 1982, Information, trade, and common knowledge, *Journal of Economic Theory* 26, 17–27.

Mullahy, J., 1986, Specification and testing of some modified count data models, *Journal of Econometrics* 33, 341–365.

Mundaca, B. G., 2000, The effect of interventions on realignment, *Journal of International Financial Markets, Institutions & Money* 10, 323–347.

Nagin, D. S., and K. C. Land, 1993, Age, criminal careers, and population heterogeneity: Specification and estimation of a nonparametric, mixed Poisson model, *Criminology* 31, 327–362.

Nelson, D. B., 1991, Conditional heteroskedasticity in asset returns: A new approach, *Econometrica* 59, 347–370.

Newey, W. K., 1985, Maximum likelihood specification testing and conditional moment tests, *Econometrica* 53, 1047–1070.

NYSE, 1996, *New York Stock Exchange Fact Book 1996*. (New York Stock Exchange New York).

Odders-White, E. R., 2000, On the occurence and consequences of inaccurate trade classification, *Journal of Financial Markets* 3, 259–286.

O'Hara, M., 1995, *Market microstructure theory*. (Basil Blackwell Oxford).

O'Hara, M., 1999, Making market microstructure matter, *Financial Management* 28, 83–90.

Perron, P., 1989, The great crash, the oil price shock and the unit root hypothesis, *Econometrica* 57, 1361–1401.

Quandt, R. E., 1958, The estimation of the parameters of a linear regression system obeying two separate regimes, *Journal of the American Statistical Association* 53, 873–880.

Quandt, R. E., 1972, A new approach to estimating switching regressions, *Journal of the American Statistical Association* 67, 306–310.

Quandt, R. E., and J. B. Ramsey, 1978, Estimating mixtures of normal distributions and switching regressions, *Journal of the American Statistical Association* 73, 730–738.

Raj, B., 2002, Asymmetry of business cycles: The Markov-switching approach, in A. Ullah, A. Wan, and A. Chaturvedi, eds.: *Handbook of Applied Econometrics and Statistical Inference* (Marcel Dekker, New York ).

Redner, R. A., and H. F. Walker, 1985, Mixture densities, maximum likelihood and the EM algorithm, *SIAM Review* 26, 195–239.

Richard, J.-F., 1980, Models with several regimes and changes in exogenity, *Review of Economic Studies* 47, 1–20.

Rochet, J.-C., and J.-L. Vila, 1994, Insider trading without normality, *Review of Economic Studies* 61, 131–152.

Roeder, K., K. G. Lynch, and D. S. Nagin, 1999, Modeling uncertainty in latent class membership: A case study in criminology, *Journal of the American Statistical Association* 94, 766–776.

Ross, S. M., 1996, *Stochastic processes*. (John Wiley & Sons New York) 2nd edn.

Russell, Jeffrey R., 1999, Econometric modeling of multivariate irregularly-spaced high-frequency data, Discussion Paper Graduate School of Business, University of Chicago.

Russell, Jeffrey R., and Robert F. Engle, 1999, Econometric analysis of discrete-valued irregularly-spaced financial transaction data, Discussion Paper Graduate School of Business, University of Chicago.

Ruud, P. A., 1991, Extensions of estimation methods using the EM algorithm, *Journal of Econometrics* 49, 307–341.

Saar, G., 2000, Prices and spreads in markets with information imperfections, Department of Finance Working Paper Stern School of Business.

Saar, G., 2001a, Investor uncertainty and order flow information, Department of Finance Working Paper Series No. FIN-01-063 Stern School of Business.

Saar, G., 2001b, Price impact asymmetry of block trades: An institutional trading explanation, *Review of Financial Studies* 14, 1153–1181.

Sawa, T., 1978, Information criteria for discriminating among alternative regression models, *Econometrica* 46, 1273–1291.

Schaller, H., and S. van Norden, 2002, Fads or bubbles?, *Empirical Economics* 27, 335–362.

Schlag, C., and H. Stoll, 2001, Price impacts of option volume, Working Paper J. W. Goethe University Frankfurt.

Schwartz, R. A., 1988, *Equity markets. Structure, trading, and performance.* (Harper & Row Publishers Cambridge).

Schwartz, R. A., 1991, *Reshaping the equity markets. A guide for the 1990s.* (Harper Buisness London).

Schwarz, G., 1978, Estimating the dimension of a model, *Annals of Statistics* 6, 461–464.

Seppi, D. J., 1990, Equilibrium block trading and asymmetric information, *Journal of Finance* 45, 73–94.

Shaked, M., 1980, On mixtures from exponential families, *Journal of the Royal Statistical Society, Series B* 42, 192–198.

Silverman, B. W., 1986, *Density Estimation for Statistics and Data Analysis.* (Chapman and Hall London).

Smith, D. R., 2002, Markov-switching and stochastic volatility diffusion models of short-term interest rates, *Journal of Business and Economic Statistics* 20, 183–197.

Spanos, A., 1986, *Statistical Foundations of Econometric Modelling.* (Cambridge University Press Cambridge).

Spiegel, M., and A. Subrahmanyam, 1992, Informed speculation and hedging in a noncompetitive securities market, *Review of Financial Studies* 5, 307–329.

Spulber, D. F., 1996, Market microstructure and intermediation, *Journal of Economic Perspectives* 10, 135–152.

Stoll, H. R., 2001, Market microstructure, Working Paper No. 01-16 Financial Markets Research Center, Vanderbilt University.

TAQ, 1996, *The TAQ database Version 3.3.* (New York Stock Exchange Inc. Manual).

Tauchen, G., 1985, Diagnostic testing and evaluation of maximum likelihood models, *Journal of Econometrics* 30, 415–443.

Theil, H., 1971, *Principles of Econometrics.* (John Wiley & Sons Santa Barbara).

Theissen, E., 2000, A test of the accuracy of the Lee/Ready trade classification algorithm, Working Paper, University of Frankfurt.

Timmermann, A., 2000, Moments of Markov switching models, *Journal of Econometrics* 96, 75–111.

Titterington, D. M., A. F. M. Smith, and U. E. Makov, 1985, *Statistcal analysis of finite mixture distributions.* (John Wiley & Sons Chichester).

Trivedi, P. K., and P. Deb, 1997, Demand for medical care by the elderly: A finite mixture approach, *Journal of Applied Econometrics* 12, 313–336.

Turner, C. M., R. Startz, and C. R. Nelson, 1989, A Markov model of heteroscedasticity, risk, and learning in the stock market, *Journal of Financial Economics* 25, 3–22.

Verrecchia, R. E., 1982, Information acquisition in a noisy rational expectations economy, *Econometrica* 50, 1415–1430.

Wang, P., I. M. Cockburn, and M. L. Puterman, 1998, Analysis of patent data - A mixed-Poisson-regression approach, *Journal of Business & Economic Statistics* 16, 27–41.

Wang, P., M. L. Puterman, I. M. Cockburn, and N. Le, 1996, Mixed Poisson regression models with covariate dependent rates, *Biometrics* 52, 381–400.

Wedel, M., W. S. Desarbo, J. R. Bult, and V. Ramaswamy, 1993, A latent class Poisson regression model for heterogenous count data, *Journal of Applied Econometrics* 8, 397–411.

Weston, J. P., 2001, Information, liquidity, and noise, Working Paper Rice University.

White, H., 1982, Maximum likelihood estimation of misspecified models, *Econometrica* 50, 1–25.

Whittemore, A. S., and G. Gong, 1991, Poisson regression with misclassified counts: Application to cervical cancer mortality rates, *Applied Statistics* 40, 81–93.

Wu, C. F. J., 1983, On the convergence properties of the EM algorithm, *The Annals of Statistics* 11, 95–103.

Zeger, S. L., and B. Qaqish, 1988, Markov regression models for time series: A quasi-likelihood approach, *Biometrics* 44, 1019–1031.

# List of Figures

3.1 The Easley, Kiefer, O'Hara and Paperman (1996) model. . . . . . . 25
3.2 The Easley, Kiefer and O'Hara (1997b) model without trade
size effects. . . . . . . . . . . . . . . . . . . . . . . . . . . . . . . . . . . . . 28
3.3 The Easley, Kiefer and O'Hara (1997b) with trade size effects. . 30
3.4 The Easley, Kiefer and O'Hara (1997a) model. . . . . . . . . . . . . 31
3.5 The Weston (2001) model. . . . . . . . . . . . . . . . . . . . . . . . . . . 33
3.6 The Easley, O'Hara and Saar (2001) model. . . . . . . . . . . . . . . 35
3.7 The Brown, Thomson and Walsh (1999) model. . . . . . . . . . . . . 36
3.8 The Grammig, Schiereck and Theissen (2001) model. . . . . . . . . 38
3.9 The Easley, Kiefer and O'Hara (1996) model. . . . . . . . . . . . . . 39
3.10 The Easley, O'Hara and Srinivas (1998) model. . . . . . . . . . . . . 41

4.1 Trades of IBM shares . . . . . . . . . . . . . . . . . . . . . . . . . . . . . . 62
4.2 Structure of the *ignorant agnostic* trading model. . . . . . . . . . . . 64
4.3 Structure of the generalized sequential trade model. . . . . . . . . . 68
4.4 Structure of the Markov switching sequential trade model. . . . . . 77

5.1 Estimates of the ACF . . . . . . . . . . . . . . . . . . . . . . . . . . . . . . 109
5.2 Seasonal components . . . . . . . . . . . . . . . . . . . . . . . . . . . . . . 110
5.3 Estimates of the distribution . . . . . . . . . . . . . . . . . . . . . . . . . . 111
5.4 Unconditional forecasts . . . . . . . . . . . . . . . . . . . . . . . . . . . . . 128
5.5 ACF of Pearson residuals . . . . . . . . . . . . . . . . . . . . . . . . . . . . 132
5.6 Residual function . . . . . . . . . . . . . . . . . . . . . . . . . . . . . . . . . . 133
5.7 Empirical and estimated density . . . . . . . . . . . . . . . . . . . . . . . . 134
5.8 Smoothed regime probabilities . . . . . . . . . . . . . . . . . . . . . . . . 138
5.9 Regime specific forecasts . . . . . . . . . . . . . . . . . . . . . . . . . . . . 141

A.1 A simulated Poisson process . . . . . . . . . . . . . . . . . . . . . . . . . . 151
A.2 Plot of the Poisson density . . . . . . . . . . . . . . . . . . . . . . . . . . . 162
A.3 Plot of the negative binomial density . . . . . . . . . . . . . . . . . . . . 163

# List of Tables

3.1   Empirical research using sequential trade models . . . . . . . . . . . . 55

5.1   Summary statistics for the TAQ data . . . . . . . . . . . . . . . . . . . . . . 95
5.2   Classification of trades . . . . . . . . . . . . . . . . . . . . . . . . . . . . . . . . . 105
5.3   Descriptive statistics . . . . . . . . . . . . . . . . . . . . . . . . . . . . . . . . . . . 106
5.4   Model selection BA . . . . . . . . . . . . . . . . . . . . . . . . . . . . . . . . . . . . 115
5.5   Model selection DIS . . . . . . . . . . . . . . . . . . . . . . . . . . . . . . . . . . . 116
5.6   Model selection IBM . . . . . . . . . . . . . . . . . . . . . . . . . . . . . . . . . . . 117
5.7   Model selection KO . . . . . . . . . . . . . . . . . . . . . . . . . . . . . . . . . . . . 118
5.8   Model selection XON . . . . . . . . . . . . . . . . . . . . . . . . . . . . . . . . . . 119
5.9   Parameter estimates BA . . . . . . . . . . . . . . . . . . . . . . . . . . . . . . . 123
5.10 Parameter estimates DIS . . . . . . . . . . . . . . . . . . . . . . . . . . . . . . 124
5.11 Parameter estimates IBM . . . . . . . . . . . . . . . . . . . . . . . . . . . . . 125
5.12 Parameter estimates KO . . . . . . . . . . . . . . . . . . . . . . . . . . . . . . 126
5.13 Parameter estimates XON . . . . . . . . . . . . . . . . . . . . . . . . . . . . . 127
5.14 Properties of the regimes . . . . . . . . . . . . . . . . . . . . . . . . . . . . . . 136
5.15 Results of parameter restriction tests . . . . . . . . . . . . . . . . . . . . 143

Druck und Bindung: Strauss GmbH, Mörlenbach

# Lecture Notes in Economics and Mathematical Systems

For information about Vols. 1–444
please contact your bookseller or Springer-Verlag

Vol. 445: A. H. Christer, S. Osaki, L. C. Thomas (Eds.), Stochastic Modelling in Innovative Manufactoring. X, 361 pages. 1997.

Vol. 446: G. Dhaene, Encompassing. X, 160 pages. 1997.

Vol. 447: A. Artale, Rings in Auctions. X, 172 pages. 1997.

Vol. 448: G. Fandel, T. Gal (Eds.), Multiple Criteria Decision Making. XII, 678 pages. 1997.

Vol. 449: F. Fang, M. Sanglier (Eds.), Complexity and Self-Organization in Social and Economic Systems. IX, 317 pages, 1997.

Vol. 450: P. M. Pardalos, D. W. Hearn, W. W. Hager, (Eds.), Network Optimization. VIII, 485 pages, 1997.

Vol. 451: M. Salge, Rational Bubbles. Theoretical Basis, Economic Relevance, and Empirical Evidence with a Special Emphasis on the German Stock Market. IX, 265 pages. 1997.

Vol. 452: P. Gritzmann, R. Horst, E. Sachs, R. Tichatschke (Eds.), Recent Advances in Optimization. VIII, 379 pages. 1997.

Vol. 453: A. S. Tangian, J. Gruber (Eds.), Constructing Scalar-Valued Objective Functions. VIII, 298 pages. 1997.

Vol. 454: H.-M. Krolzig, Markov-Switching Vector Autoregressions. XIV, 358 pages. 1997.

Vol. 455: R. Caballero, F. Ruiz, R. E. Steuer (Eds.), Advances in Multiple Objective and Goal Programming. VIII, 391 pages. 1997.

Vol. 456: R. Conte, R. Hegselmann, P. Terna (Eds.), Simulating Social Phenomena. VIII, 536 pages. 1997.

Vol. 457: C. Hsu, Volume and the Nonlinear Dynamics of Stock Returns. VIII, 133 pages. 1998.

Vol. 458: K. Marti, P. Kall (Eds.), Stochastic Programming Methods and Technical Applications. X, 437 pages. 1998.

Vol. 459: H. K. Ryu, D. J. Slottje, Measuring Trends in U.S. Income Inequality. XI, 195 pages. 1998.

Vol. 460: B. Fleischmann, J. A. E. E. van Nunen, M. G. Speranza, P. Stähly, Advances in Distribution Logistic. XI, 535 pages. 1998.

Vol. 461: U. Schmidt, Axiomatic Utility Theory under Risk. XV, 201 pages. 1998.

Vol. 462: L. von Auer, Dynamic Preferences, Choice Mechanisms, and Welfare. XII, 226 pages. 1998.

Vol. 463: G. Abraham-Frois (Ed.), Non-Linear Dynamics and Endogenous Cycles. VI, 204 pages. 1998.

Vol. 464: A. Aulin, The Impact of Science on Economic Growth and its Cycles. IX, 204 pages. 1998.

Vol. 465: T. J. Stewart, R. C. van den Honert (Eds.), Trends in Multicriteria Decision Making. X, 448 pages. 1998.

Vol. 466: A. Sadrieh, The Alternating Double Auction Market. VII, 350 pages. 1998.

Vol. 467: H. Hennig-Schmidt, Bargaining in a Video Experiment. Determinants of Boundedly Rational Behavior. XII, 221 pages. 1999.

Vol. 468: A. Ziegler, A Game Theory Analysis of Options. XIV, 145 pages. 1999.

Vol. 469: M. P. Vogel, Environmental Kuznets Curves. XIII, 197 pages. 1999.

Vol. 470: M. Ammann, Pricing Derivative Credit Risk. XII, 228 pages. 1999.

Vol. 471: N. H. M. Wilson (Ed.), Computer-Aided Transit Scheduling. XI, 444 pages. 1999.

Vol. 472: J.-R. Tyran, Money Illusion and Strategic Complementarity as Causes of Monetary Non-Neutrality. X, 228 pages. 1999.

Vol. 473: S. Helber, Performance Analysis of Flow Lines with Non-Linear Flow of Material. IX, 280 pages. 1999.

Vol. 474: U. Schwalbe, The Core of Economies with Asymmetric Information. IX, 141 pages. 1999.

Vol. 475: L. Kaas, Dynamic Macroeconomics with Imperfect Competition. XI, 155 pages. 1999.

Vol. 476: R. Demel, Fiscal Policy, Public Debt and the Term Structure of Interest Rates. X, 279 pages. 1999.

Vol. 477: M. Théra, R. Tichatschke (Eds.), Ill-posed Variational Problems and Regularization Techniques. VIII, 274 pages. 1999.

Vol. 478: S. Hartmann, Project Scheduling under Limited Resources. XII, 221 pages. 1999.

Vol. 479: L. v. Thadden, Money, Inflation, and Capital Formation. IX, 192 pages. 1999.

Vol. 480: M. Grazia Speranza, P. Stähly (Eds.), New Trends in Distribution Logistics. X, 336 pages. 1999.

Vol. 481: V. H. Nguyen, J. J. Strodiot, P. Tossings (Eds.). Optimation. IX, 498 pages. 2000.

Vol. 482: W. B. Zhang, A Theory of International Trade. XI, 192 pages. 2000.

Vol. 483: M. Königstein, Equity, Efficiency and Evolutionary Stability in Bargaining Games with Joint Production. XII, 197 pages. 2000.

Vol. 484: D. D. Gatti, M. Gallegati, A. Kirman, Interaction and Market Structure. VI, 298 pages. 2000.

Vol. 485: A. Garnaev, Search Games and Other Applications of Game Theory. VIII, 145 pages. 2000.

Vol. 486: M. Neugart, Nonlinear Labor Market Dynamics. X, 175 pages. 2000.

Vol. 487: Y. Y. Haimes, R. E. Steuer (Eds.), Research and Practice in Multiple Criteria Decision Making. XVII, 553 pages. 2000.

Vol. 488: B. Schmolck, Ommitted Variable Tests and Dynamic Specification. X, 144 pages. 2000.

Vol. 489: T. Steger, Transitional Dynamics and Economic Growth in Developing Countries. VIII, 151 pages. 2000.

Vol. 490: S. Minner, Strategic Safety Stocks in Supply Chains. XI, 214 pages. 2000.

Vol. 491: M. Ehrgott, Multicriteria Optimization. VIII, 242 pages. 2000.

Vol. 492: T. Phan Huy, Constraint Propagation in Flexible Manufacturing. IX, 258 pages. 2000.

Vol. 493: J. Zhu, Modular Pricing of Options. X, 170 pages. 2000.

Vol. 494: D. Franzen, Design of Master Agreements for OTC Derivatives. VIII, 175 pages. 2001.

Vol. 495: I Konnov, Combined Relaxation Methods for Variational Inequalities. XI, 181 pages. 2001.

Vol. 496: P. Weiß, Unemployment in Open Economies. XII, 226 pages. 2001.

Vol. 497: J. Inkmann, Conditional Moment Estimation of Nonlinear Equation Systems. VIII, 214 pages. 2001.

Vol. 498: M. Reutter, A Macroeconomic Model of West German Unemployment. X, 125 pages. 2001.

Vol. 499: A. Casajus, Focal Points in Framed Games. XI, 131 pages. 2001.

Vol. 500: F. Nardini, Technical Progress and Economic Growth. XVII, 191 pages. 2001.

Vol. 501: M. Fleischmann, Quantitative Models for Reverse Logistics. XI, 181 pages. 2001.

Vol. 502: N. Hadjisavvas, J. E. Martínez-Legaz, J.-P. Penot (Eds.), Generalized Convexity and Generalized Monotonicity. IX, 410 pages. 2001.

Vol. 503: A. Kirman, J.-B. Zimmermann (Eds.), Economics with Heterogenous Interacting Agents. VII, 343 pages. 2001.

Vol. 504: P.-Y. Moix (Ed.),The Measurement of Market Risk. XI, 272 pages. 2001.

Vol. 505: S. Voß, J. R. Daduna (Eds.), Computer-Aided Scheduling of Public Transport. XI, 466 pages. 2001.

Vol. 506: B. P. Kellerhals, Financial Pricing Models in Continuous Time and Kalman Filtering. XIV, 247 pages. 2001.

Vol. 507: M. Koksalan, S. Zionts, Multiple Criteria Decision Making in the New Millenium. XII, 481 pages. 2001.

Vol. 508: K. Neumann, C. Schwindt, J. Zimmermann, Project Scheduling with Time Windows and Scarce Resources. XI, 335 pages. 2002.

Vol. 509: D. Hornung, Investment, R&D, and Long-Run Growth. XVI, 194 pages. 2002.

Vol. 510: A. S. Tangian, Constructing and Applying Objective Functions. XII, 582 pages. 2002.

Vol. 511: M. Külpmann, Stock Market Overreaction and Fundamental Valuation. IX, 198 pages. 2002.

Vol. 512: W.-B. Zhang, An Economic Theory of Cities.XI, 220 pages. 2002.

Vol. 513: K. Marti, Stochastic Optimization Techniques. VIII, 364 pages. 2002.

Vol. 514: S. Wang, Y. Xia, Portfolio and Asset Pricing. XII, 200 pages. 2002.

Vol. 515: G. Heisig, Planning Stability in Material Requirements Planning System. XII, 264 pages. 2002.

Vol. 516: B. Schmid, Pricing Credit Linked Financial Instruments. X, 246 pages. 2002.

Vol. 517: H. I. Meinhardt, Cooperative Decision Making in Common Pool Situations. VIII, 205 pages. 2002.

Vol. 518: S. Napel, Bilateral Bargaining. VIII, 188 pages. 2002.

Vol. 519: A. Klose, G. Speranza, L. N. Van Wassenhove (Eds.), Quantitative Approaches to Distribution Logistics and Supply Chain Management. XIII, 421 pages. 2002.

Vol. 520: B. Glaser, Efficiency versus Sustainability in Dynamic Decision Making. IX, 252 pages. 2002.

Vol. 521: R. Cowan, N. Jonard (Eds.), Heterogenous Agents, Interactions and Economic Performance. XIV, 339 pages. 2003.

Vol. 522: C. Neff, Corporate Finance, Innovation, and Strategic Competition. IX, 218 pages. 2003.

Vol. 523: W.-B. Zhang, A Theory of Interregional Dynamics. XI, 231 pages. 2003.

Vol. 524: M. Frölich, Programme Evaluation and Treatment Choise. VIII, 191 pages. 2003.

Vol. 525:S. Spinler, Capacity Reservation for Capital-Intensive Technologies. XVI, 139 pages. 2003.

Vol. 526: C. F. Daganzo, A Theory of Supply Chains. VIII, 123 pages. 2003.

Vol. 527: C. E. Metz, Information Dissemination in Currency Crises. XI, 231 pages. 2003.

Vol. 528: R. Stolletz, Performance Analysis and Optimization of Inbound Call Centers. X, 219 pages. 2003.

Vol. 529: W. Krabs, S. W. Pickl, Analysis, Controllability and Optimization of Time-Discrete Systems and Dynamical Games. XII, 187 pages. 2003.

Vol. 530: R. Wapler, Unemployment, Market Structure and Growth. XXVII, 207 pages. 2003.

Vol. 531: M. Gallegati, A. Kirman, M. Marsili (Eds.), The Complex Dynamics of Economic Interaction. XV, 402 pages, 2004.

Vol. 532: K. Marti, Y. Ermoliev, G. Pflug (Eds.), Dynamic Stochastic Optimization. VIII, 336 pages. 2004.

Vol. 533: G. Dudek, Collaborative Planning in Supply Chains. X, 234 pages. 2004.

Vol. 534: M. Runkel, Environmental and Resource Policy for Consumer Durables. X, 197 pages. 2004.

Vol. 535: X. Gandibleux, M. Sevaux, K. Sörensen, V.T'kindt (Eds.), Metaheuristics for Multiobjective Optimisation. IX, 249 pages. 2004.

Vol. 536: R. Brüggemann, Model Reduction Methods for Vector Autoregressive Processes. X, 218 pages. 2004.

Vol. 537: A. Esser, Pricing in (In)Complete Markets. XI, 122 pages, 2004.

Vol. 538: S. Kokot, The Econometrics of Sequential Trade Models. XI, 193 pages. 2004.

Printed in the United States
99577LV00001B/91/A

9 783540 208143